ENERGY 2040

Deepak Divan • Suresh Sharma

ENERGY 2040

Aligning Innovation, Economics and Decarbonization

Deepak Divan
Center for Distributed Energy
Georgia Institute of Technology
Atlanta, GA, USA

Suresh Sharma
Center for Distributed Energy
Georgia Institute of Technology
Atlanta, GA, USA

ISBN 978-3-031-49416-1 ISBN 978-3-031-49417-8 (eBook)
https://doi.org/10.1007/978-3-031-49417-8

This Springer imprint is published by the registered company Springer Nature Switzerland AG
The registered company address is: Gewerbestrasse 11, 6330 Cham, Switzerland

Paper in this product is recyclable.

My wife Anu, for years of unquestioning support and love
– Deepak Divan, Ph.D.

My three lovely little grandsons: Grady Rao Sharma, Fionn Cramer Sharma, and Seamus Srinivas Sharma, who would hopefully enjoy a more sustainable and cleaner planet if the ongoing energy transitions go right.
– Suresh Sharma

Acknowledgments

This book and its insights are a result of many decades of working in and with the energy industry – in advanced research, academia, startups; with industry and utility leaders; with researchers, engineers, and field technicians; with regulators, lawyers, and policy wonks; with leading VCs, investors, and entrepreneurs; with global manufacturing organizations; with national labs and research institutions; and with global institutions that shape how the world moves forward. Despite this close association, it was perhaps our entrepreneurial experiences that taught us the most.

We sincerely acknowledge everyone above for the things we learned and for the numerous interactions we have had. Given that we have so many of our colleagues, mentors, staff, mentees, supporters, students, and friends who have made a difference, it is difficult to individually acknowledge everyone. We have tried to capture as much as possible in the form of bibliography, references, and underlying citations for the body of their work.

Along this journey, the most important thing was not to take the status quo, as it was presented, at face value, but to ask fundamental questions that looked beyond the narrow narrative that was being presented. For this, I (Deepak) have to thank my friend and mentor, Don Novotny, who, in response to being presented what we thought was a novel solution, never ceased to flummox me with the most simple and basic of questions – he is deeply missed.

To "poke" at what seemed to be consistent pictures and stories that were accepted by everyone became second nature, allowing us to understand at a deeper level why something was, or was not, really possible, and to answer the question of whether the solutions were ultimately scalable at reasonable cost and along a needed timeline. The repertoire of an almost infinite number of questions showed us the way forward. We became focused on making sure all the key questions were asked – as the questions continued, the answers became self-evident.

As the decarbonization, climate change, and energy access/resiliency threads started getting tangled together, we began to see this as a mission for survival of the human race, certainly with the standard of living we enjoy in the global north. Time was running out, and we could see perhaps one path that could align the forces of innovation, economics, and decarbonization. Only then, we felt, could climate change be meaningfully addressed. We hope we have shown a meaningful way forward.

The number of people and organizations to be thanked and acknowledged is far too many to include here, but a few merit special mention. First of all, the graduate students, research scholars, and staff from the Georgia Tech Center for Distributed Energy, who made all the magic happen, are an integral part of who we are and what we do. Georgia Research Alliance bought into the vision of CDE and funded us to create the technology base that can play a major part in the energy transition. Countless conversations with leadership and staff at EPRI and utilities, including our partner Southern Company, helped us to understand the issues we are seeing. Leadership from DOE and industrial leaders such as GE and Siemens, and many smaller companies provided context and color to what is happening. Conversations with dozens of leading VCs and investors provided glimpses of successes and challenges. The US National Academy of Engineering, through its Board of Energy and Environmental Systems (BEES) and the committee on "The Future of Electric Power in the United States" gave access to a unique group who continually amazes and informs us. Frank Kreikebaum and Mahesh Morjaria provided us with invaluable feedback by reviewing a draft of the manuscript. We also want to call out Joseph Benzaquen from GT-CDE, who has provided valuable feedback on many of the ideas and has helped with very compelling figures. Sean DiLeonardi is also acknowledged for help with organizing the manuscript in the early months.

Finally, this book could not have been done without the patience, love, and support from our respective families, especially our spouses – Anu for Deepak, and Lata for Suresh.

Thanks to Springer Natural for agreeing to publish this book, and for the support from the Editorial team for getting it to print and help in reaching our target audiences with what we believe can have impact on how the energy transition unfolds.

Prologue

Much has happened since early 2020 when the idea for this book was developed. Even though we were all preoccupied with an unprecedented global pandemic, it was clear to us, even then, that the two intertwined mega-issues that would survive the immediate challenges, and that we as humans had to deal with sooner rather than later, were energy and climate change. In the days since, it has become abundantly clear that, with the best of intentions, we are confused and struggling to converge on a preferred path forward. Many new technologies seem to have come from nowhere, and to have gained velocity as they are adopted across the world. But it is not clear that the much talked about "energy transition" is well understood, and many issues are certainly not resolved. Widely varying views of what is happening paint either optimistic utopian scenarios or dark dystopian scenarios, both of which are probably incorrect, at least at some level. The deeply technical nature of this transformation can make it difficult for the common man to understand the nuances of what is driving the rapid change, and what the trade-offs are. Oversimplification of the issues can lead to a misunderstanding of the challenges, and what should be done about them. We are becoming a society that is deeply divided on all manner of issues, including in the critical areas of climate change and energy.

During the pandemic, with some shelter from the day-to-day emergencies that otherwise dominate our daily lives, we (the authors) had time for introspection and could see that despite our many seemingly irreconcilable differences at a societal level, we were also all united by a few common aspirations. Rather than vilifying people with a different point of view, it became important that we attempt to understand the underlying reasons behind why and how reasonable people could hold such widely differing positions. Until we deeply understood everyone's perspectives, we felt we would not be able to develop a unifying conversation that would be understood by everyone.

Massive and fast-moving changes in many energy-related areas are either already upon us or seem likely over the next 10–20 years. Disruption on such a global scale produces winners and losers. Who decides who wins and who loses? Who pays for any additional costs that may come from addressing climate change issues? In the face of such a moral dilemma, how can we converge on an acceptable path forward? It seemed critical to us that we understand and be guided by history, so we did not repeat the same mistakes. But we were also concerned that a massive and fundamental paradigm shift may already be underway, making it very challenging to blindly apply the lessons of history. The institutions and policies that have guided us over the last fifty, even a hundred, years have worked well in the past, but now seem challenged to cope with the new intersecting realities of multiple rapidly changing fronts that are seemingly moving along uncoordinated trajectories. This fast-evolving story is global and complex, making it challenging to unpack the fundamental drivers.

Both climate and energy are top-of-mind issues, with a vast and still growing stream of books and publications written by international and national committees, think tanks, academics, venture capitalists, policy wonks, and major authors. Many of the energy related books we read were wonderful, exciting our imagination, closing gaps in our understanding, and outlining the complex interplay of history, science, politics, and economics that underpinned the evolution of this complex field. Books on climate were scientifically compelling, laying out the data and consequences of inaction in a manner that could not be contested. There was no lack of prescriptive action from both sides, all to achieve the objectives that they espoused. But to our mind, the division between energy and climate change represented a rift in the discourse that has in turn divided our communities and even the world. A path to decarbonization or climate impact mitigation is often viewed as anti-economic, while the need to ensure the economic welfare of our families is seen as being against humanity.

What was missing in the discussion, we felt, was a holistic, and yet grounded, view of energy, its growth and impact on climate, and an understanding of how innovation, fast-moving technologies, and the human spirit can act as agents of change. It seemed to us that it was possible to create strong alignment between the forces of economics and climate change, and when this was achieved, there was no limit to how fast we could adopt and adapt to meet both goals. With simultaneous rapid change on many fronts comes uncertainty in terms of outcomes – which makes us uncomfortable committing to major initiatives into the unknown. Rather than freezing us in our tracks, we can use that uncertainty to create a new twenty-first-century ethos, where we create solutions that are democratized, flexible, interoperable, equitable, and scalable. A rigid scaffold, on which all development and progress occurs along a prescribed 20-year plan, is not how things evolve in nature – but that is how we have viewed and wanted our own world to be: rigid and predictable. We want familiarity and for our lives to remain undisrupted. Maybe fast-evolving, twenty-first-century technologies can offer a new approach to building a future world that meets human needs but is also designed to preserve our planet for future generations.

This body of work has been inspired and informed by our own experiences, our market research, and more recently, lessons learned from an experiment we led at the Center for Distributed Energy at the Georgia Institute of Technology in Atlanta, Georgia, USA. Our goal was to develop a coherent view of the current energy transition, especially where it intersected with the electricity grid, which we felt would be an even more important part of a future energy infrastructure. We were a little concerned about what we, as two technology and innovation practitioners, could contribute to this widely researched and published area. We wanted to see if we could develop an actionable pathway to evolve a future energy infrastructure that is aligned with game-changing transformations going on in energy innovations, economics of energy, and its impact on climate through decarbonization.

It may be appropriate to briefly share our backgrounds, and to see how our experiences helped shape the context and content of this book. Deepak has spent 40 years in academia and in founding and running startups, and currently serves as Professor, GRA Eminent Scholar, and the Founding Director of the Center for Distributed Energy at the Georgia Institute of Technology. He is an elected member of the US National Academy of Engineering and has been active in research, working closely with industry to develop and translate technologies to market in many of the related technology areas. As a serial entrepreneur, he has worked to understand how new disruptive technologies make their way to market, and has raised money from leading VCs, building teams, launching new products, and grappling with why the obvious pathways were not so obvious in the regulated utility world. He has served as an advisor to organizations such as EPRI and GE Research in the early days, and has served on the NASEM Board on Energy and Environmental Systems. He has been an invited speaker at many global meetings, including at COP 22 in Marrakesh in 2016, and at the UN Global Solutions Summit in 2023. He was also a member of the recent NASEM Committee on "The Future of Electric Power in the US" and contributed to the influential consensus report that was issued in 2021. Over 45 years of working within IEEE societies, including in leadership positions, has given him visibility and familiarity with cutting-edge technologies being developed, not only in the USA, but across the globe.

Suresh, similarly, has had a wide range of experiences in industry and business, starting his early career as a Naval Aviation Officer in India and the UK, subsequently working in technology and business leadership roles for many years at GE Energy, both in the USA and other global locations. He led the global development and implementation of several new technologies for the energy, aviation, and aerospace industry. The coauthors have learned a lot through a shared entrepreneurial experience at Innovolt, and then over the past seven years at the Center for Distributed Energy at Georgia Tech, where Suresh held the unique position of Entrepreneur in Residence (EIR). Suresh's passion for understanding the intricacies of innovation and entrepreneurship have led to many books on diverse topics, including *Industrializing Innovation* (2019) and *The 3rd American Dream (2014)*, books that intersect heavily with the issues under discussion here.

We hope you enjoy reading the book as much as we have enjoyed writing it.

Atlanta, GA, USA
October 2023

Deepak Divan
Suresh Sharma

Acronyms, Abbreviations, and Scientific Units

Technical Acronyms

AGC	Automatic Generator Control
AMI	Automated Metering Infrastructure
CAES	Compressed Air Energy Storage
CCGT	Combined Cycle Gas Turbine
CCS	Carbon Capture and Sequestration
DER	Distributed Energy Resources
DERMS	DER Management System
DFIG	Doubly Fed Induction Generator
DMS	Distribution Management System
DSP	Digital Signal Processor
EMS	Energy Management System
EV	Electric Vehicle
FIT	Feed in Tariff
FPGA	Field Programmable Gate Array
GHG	Green House Gases
HEV	Hybrid Electric Vehicle
HILF	High Impact Low Frequency
HVDC	High Voltage DC
IGBT	Insulated Gate Bipolar Transistor
IRP	Integrated Resource Plan
LCOE	Levelized Cost of Energy
LDC	Least Developed Country
LDES	Long Duration Energy Storage
LMP	Locational Marginal Pricing
MOSFET	Metal Oxide Semiconductor Field Effect Transistor
MPPT	Maximum Peak Power Tracking
MTDC	Multi-Terminal DC
MVP	Minimum Viable Prototype
PHEV	Plug-in Hybrid Electric Vehicle
PMU	Phasor Measurement Unit
PV	Photovoltaic

RPS	Renewable Portfolio Standard
RTMR	Real Time Must Run
SAIDI	System Average Interruption Duration Index
SCADA	Supervisory Control and Data Acquisition
SCED	Security Constrained Economic Dispatch
SMES	Superconducting Magnetic Energy Storage
UPS	Uninterruptible Power Supply
V2G	Vehicle to Grid
V2H	Vehicle to Home

Institutions/Non-technical

DOE	Department of Energy
EPACT 1992	Energy Policy Act 1992
EPRI	Electric Power Research Institute
COPXX	Conference of Parties
FERC	Federal Energy Regulatory Commission
HDI	Human Development Index
IEA	International Energy Agency
IEEE	Institution of Electrical and Electronics Engineers
IFC	International Finance Corporation
IOU	Investor-Owned Utility
IPCC	Intergovernmental Panel on Climate Change
IPP	Independent Power Producer
ISO	Independent System Operator
NAE	National Academy of Engineering
NARUC	National Association of Regulatory Utility Commissioners
NASA	National Aeronautics and Space Administration
NASEM	National Academy of Science Engineering and Medicine
NERC	North American Electric Reliability Corporation
NIMBY	Not In My Back Yard
NRECA	National Rural Electric Cooperative Association
NREL	National Renewable Energy Laboratory
NSF	National Science Foundation
PUHCA	Public Utility Holding Company Act
PURPA	Public Utilities Regulatory Policies Act
ONR	Office of Naval Research
OPEC	Organization of Petroleum Exporting Countries
RTO	Regional Transmission Operator
SGIG	Smart Grid Investment Grant
SPAC	Special Purpose Acquisition Company
US-EIA	US Energy Information Administration

Key Scientific Units

BTU – British Thermal Unit: Unit of energy used as a measure of heat – 1 BTU = 1055 Joules
Quad – Unit of energy – 1 Quad = 10^{15} BTU, or 293 TWh or 33 GW-years

Electrical Power and Energy

Power represents work done, in this case using electricity.

Watt – Unit of power. 1 watt (W) = 1 volt x 1 Ampere

(representative consumption levels – LED light ~5 watts, small US home ~ 10,000 W, Boeing 737 aircraft ~ 25,000,000 watts at takeoff, Chicago peak demand ~ 25,000,000,000 watts)

kW – Kilowatt, equals 1000 W
MW – Megawatt, equals 1000 kW
GW – Gigawatt, equals 1000 MW
TW – Terawatt, equals 1000 GW

Energy is the net work done by the application of a certain level of power for a specified time.

Units of electrical energy are in (Power x Hours) – Wh, kWh, MWh, GWh, and TWh.

Scale of Energy Consumption

US annual energy consumed ~100 Quads
US peak electricity generation ~ 1000 GW
US annual electrical energy consumption ~ 4000 TWh

Contents

1 Energy and Society: At a Tipping Point

The World in Crisis

The world seems to be poised on the edge of a precipice, careening toward an unsustainable level of carbon emissions that is accelerating the pace of climate change, a factor that imperils our economic well-being and our very existence. With the best of intentions, twentieth century technology, policies, and financial instruments have been unable to address the challenges, setting up the mitigation of climate change as a Faustian bargain between economics and sustainability. Even though there is wide scientific consensus on the severe impact that climate change will have globally, the motivation for meaningful and immediate action has been limited. This is because, up until now, there has been no direct near-term incentive for decision makers to fix the issue, and because any future cost and impact of not taking any action today can be socialized and transferred to future generations (who are not here to fight for their rights).

As intelligent, educated, and rational beings, we understand the urgency of the situation and believe we would do anything to solve the problem – except perhaps change our own behavior and imperil our personal financial future, especially when we personally did not cause the problem. Or perhaps, we worry that the different groups we identify with – our nations, professions, generations, etc. – did not cause the problem, and we have the right to get to a level of prosperity that the others are at, before we start rectifying our ways. Or maybe we are fossil fuel producers and feel that the link between climate change and carbon is tenuous at best (after all the world has gone through many climate change cycles in the past, cycles that were not linked with anthropogenic carbon emissions), and that it would be foolish to compromise our livelihood and economic growth in response to uncertain and unpredictable scenarios. Or perhaps, we have built global businesses that have brought prosperity to the world, and feel that we need to protect our investors, workers, and customers, and that there is no way that unproven new disruptive solutions could ever deliver on their promise in a timely manner without causing economic havoc.

D. Divan, S. Sharma, *ENERGY 2040*,
https://doi.org/10.1007/978-3-031-49417-8_1

And we would all be right – in our own way. After all we have spent decades building our personal world view and zone of expertise and feel that we have an economically and morally defensible perspective. As a corollary, we believe others are perhaps not seeing the important elements in this picture. We believe strongly in our point of view and regard as fact only those points that support our opinions and which are then amplified via the internet, social media, and print into waves that ignite passion at all levels. Vilifying our partners along this journey to sustainability is perhaps the worst way of aligning our objectives and achieving progress. Yet, that is precisely where we frequently find ourselves, in opposite camps: haves versus have-nots; developed versus emerging nations; advocates for carbon mitigation versus supporters of economic growth; academic idealists versus practical doers. In such a deeply divided world, how do we get to alignment, to convergence?

Since 1992, when the IPCC report raised the early alarms on anthropogenic carbon emissions and its potential impact on humanity (the planet will do just fine, the people on it may not), we as society have struggled with how to respond. Each successive global summit (with the most recent COP 27 in Cairo and COP 28 in Dubai) has tried to bring all nations together – very challenging given the multitude of issues to be considered, and a highly turbulent political and policy-making process that swings like a yo-yo based on who is in power. Decarbonization goals are set every time, but with poor compliance mechanisms and no real accountability at a national level (in any case, change of political realities can completely alter the near-term strategy, as we saw for the USA during the transitions from the Obama to the Trump and then to the Biden administrations). While this political process is critically important and we are undoubtedly making progress, there is reason to be concerned that the process will not get us to zero carbon emissions in a timely manner.

For those of us old enough to have lived through the second half of the twentieth century, the world still seems very familiar: our homes, cars, and families, the way we live and travel, what we learned in school – the core fundamentals of our lives – remain steady. But something also feels very different. We know that digital technology and electronics have altered how we work, drive, play, communicate, and shop. But our material world does not seem to have been highly disturbed (although changes are now visible at the edges). For many, even as fast paced changes have occurred in the "digital" world, the "real" world seems to be changing at a more measured manageable pace. This is comforting, because no one wants accelerating runaway change to create chaos and uncertainty in their life. Yet we are also worried that rapid change seems to be coming at us from every direction, change that we did not see coming, even a short 10 years ago. Whether we want to acknowledge it or not, we know that the changes in climate simultaneously occurring across the globe cannot all be coincidence. But if the climate is actually changing because of human activity, the problem seems so big and complex that it calls into question whether we can do anything meaningful in terms of impact over the next 20 years, a period that climate scientists have indicated is critical if runaway global warming is to be avoided.

At a fundamental level, energy is at the heart of this challenge – directly accounting for as much as 75% of all carbon emissions when electricity generation, heating and cooling, industry, and transportation are all considered. We cannot reduce carbon emissions without solving the energy problem. But the world will not function without access to as much affordable, reliable, and safe energy as it needs – that must be a priority. To get to zero carbon emissions, we need to either stop emitting CO_2 (Carbon Dioxide) and other greenhouse gases (GHG) or we need to extract them from the atmosphere. We believe this needs to happen sooner rather than later, say by 2040!

On the other hand, the level of improvement in human standard of living that access to energy and technology has wrought over the last 100+ years is unbelievable and sets a golden standard that should be preserved. Despite all our accomplishments, and there are too many to count, 700 million people still live off-grid, and 3 billion live with energy poverty so extreme that it impacts their ability to earn a livelihood. Clearly the benefits of energy as available today are not equitably distributed, and we must ensure that any future we move toward resolves rather than exacerbates this inequity. We strongly believe that most people are not against energy being clean and sustainable. Their main concern is that they do not want to go back to an era where energy costs are high and reliability is poor, compromising economic gains that have been made over the last century.

This is the conundrum that is baffling us. Can we have our cake and eat it, too? Can we sustain our economies and move toward global prosperity, while still meeting decarbonization goals and better positioning ourselves to manage the impact of climate change? Are there one or two primary levers that can give us substantial impact and get us most of the way there? Can this process be equitable, bringing the economically disadvantaged to parity with the rest of the world? Or are we condemned to a repeat of history – a battle between haves and have-nots, a story of dominance by the few with geopolitical or economic clout, and increasing disparity in a resource constrained world?

Energy 2040 is about taking a fresh look at the situation we find ourselves in today, identifying the contributing factors that got us here, and understanding where we may be headed. To do that, we need to explore and harmonize a complex, multifaceted story of energy involving topics as varied as the economics of sustainability, the process of scaling new technologies, academic research, government regulation, digitalization, and energy equity – to name but a few. Many books being published today have considered the problem of energy more narrowly, including only one or two of these topics. *Energy 2040* offers a more grounded and yet holistic and comprehensive look at our energy past, and present, in order to outline potential paths to an energy future that is both sustainable and economically viable.

It's All About Energy

Energy consumption and human development are deeply interconnected. Based on UN Sustainable Development Goals (SDG), energy directly links and enables virtually every key societal objective [1]. Similarly, Fig. 1.1 shows the linkage between energy consumption and a "human development index" (HDI) score, showing that

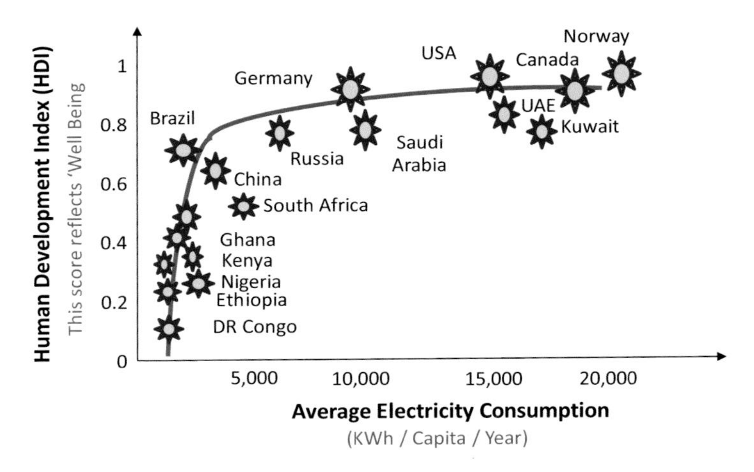

Fig. 1.1 Human development index versus. Energy consumption per capita for a set of different countries. (Representative of Connection between Energy and Well-Being)

an increase in HDI requires a disproportionate increase in energy consumption [2] Energy is clearly critically important for human development and growth.

At the global level, total energy consumed in 2020 was around 550 Quads, or 5.5 times US consumption, with the biggest consumer of energy being China (1 Quad is 10^{15} BTU or 293 million megawatt-hours, sufficient to meet all US energy needs for 3.5 days).

As detailed in Fig. 1.2, in 2019, the USA consumed over 100 Quads of energy. Of that, 80.2 Quads of primary energy came from fossil fuels, which in turn generated 75% of our anthropogenic carbon emissions. A total of 67.5 Quads of energy was rejected as waste heat due to inefficient energy conversion processes, most of which was extrinsic to the planet's natural energy balance and further added to global warming, especially at a regional level [3].

This shows that for every dollar we spend to extract, process, deliver, and convert the primary source to thermal energy, we then proceed to throw away 78% of that energy, and only use 22% for useful work [3]. By way of contrast, losses in the electrical chain, from generation to load, are very low – typically aggregating 7–10% total.

So, from an economic perspective, we all need plentiful, abundant energy to power our needs – from lighting to electronics to transportation to space heating, cooling, and industry. True, our ancestors lived in the dark and did not have any modern conveniences, but by the end of the twentieth century we had generally resolved that situation, at least in the developed countries. As we moved out of the twentieth century, it seemed that we were at a stable place in terms of energy, and that disturbing the status quo with unproven technologies would be problematic and very expensive. Emerging markets were focused on economic growth and were building coal plants as fast as they could to fuel their economies. The International Energy Authority, as well as every major energy company and gas and electric

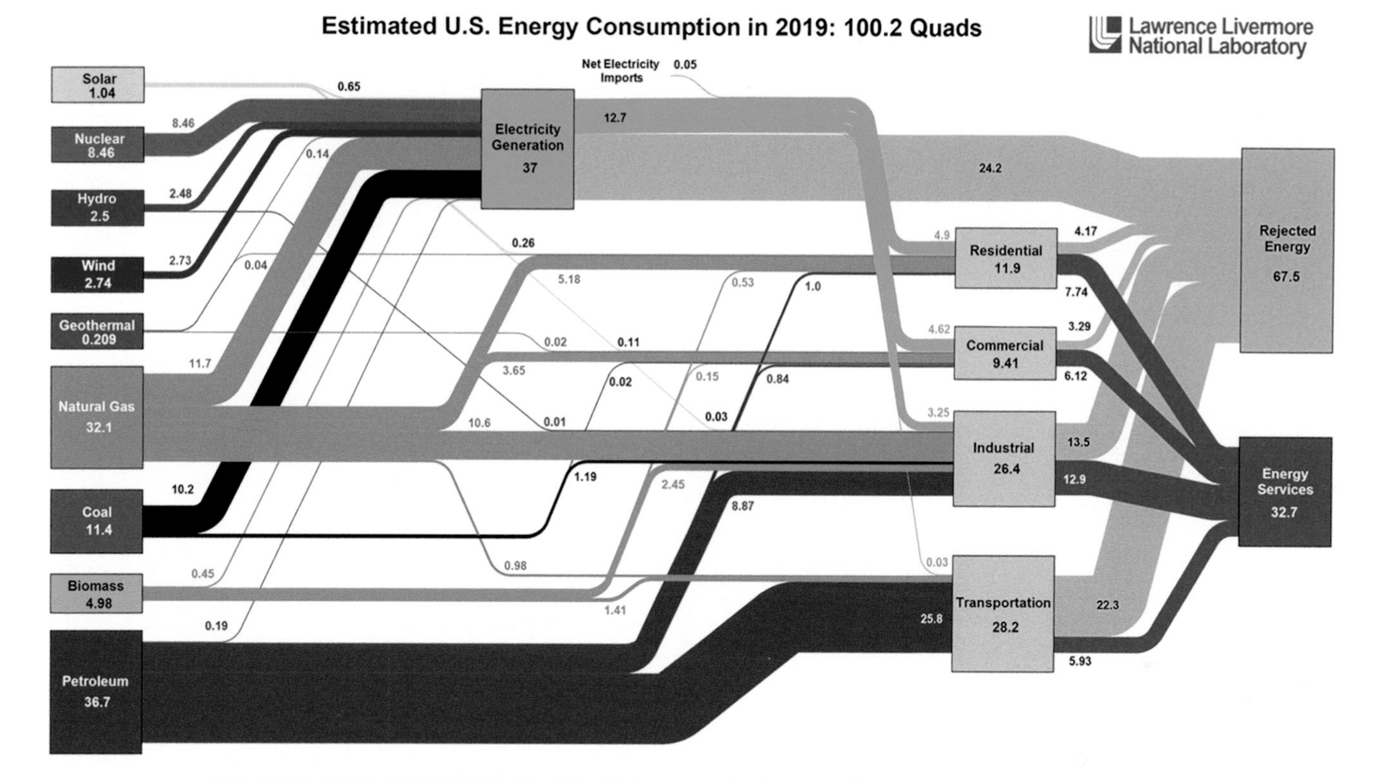

Fig. 1.2 78% of energy from fossil fuels is rejected as waste heat

utility in the world, thought renewable energy was aspirational and would only gradually increase in penetration, driven mainly by policy and incentives (not by economics). Major automotive companies were building better cars and were looking to expand markets globally, and the focus was on the internal combustion engine – after all GM had tried electric cars with the EV1 in the 1990s and the noble initiative had failed! Everybody was aligned as to where we were going!

The first 20 years of the twenty-first century have completely changed our perspective. Wind and solar are now cheaper than coal and natural gas. Electric vehicles are ramping up at an unbelievable pace, with every major automotive manufacturer committed to electrification. Where did this come from, and why did we not see it coming? The good news is that most major countries, large corporations, and many energy companies (including electric utilities) are finally embracing a goal of 50% carbon emissions reduction by 2030 and 100% carbon emissions reduction by 2050. This alignment, at least in terms of high-level goals, is very important. But does it mean we are finally over the hill and now it is simply a matter of implementing the vision?

The scale at which this change is needed (we are talking major disruption and societal transformation here), and the pace at which it has to happen (in the next 20–30 years!) is daunting. To achieve this goal, it understandably feels likely that we do not have the time to develop and adopt risky new solutions to hit 2030 and 2050 targets. Many feel that we must move fast and implement proven solutions that we have already developed. But do we fully understand the new world we are creating and the new questions that need to be asked? Will the new world look like an incremental evolution from our existing world, or will it be different in fundamental ways? Will the old rules still apply, and if not, do we know what the new rules are? Will solutions developed with yesterday's questions and technology solve tomorrow's problems, especially if we do not even know what the new issues are? Who are the experts who can guide us in this transformation? Given that predictions made by all levels of experts over the last 20+ years have been consistently and spectacularly wrong, how can we rely on the guidance of experts (who typically became experts on yesterday's systems and technologies) on how an unknown and possibly unknowable future will evolve? How many years will this change take, and what will the economic consequences of this change be? How do we get to this new world if we don't even know where we are going? We can surely try to prepare for what we know we don't know, but how can one anticipate and prepare for something that we don't know that we don't know?

Apologies to our gentle reader for posing all these difficult questions right at the start of this book. But we feel that many reasonable people are probably struggling with some version of these questions, even as we try to cope, both professionally and personally, with the pace at which our own lives are being transformed. That these issues can have severe impact on our pocketbooks and the safety of our families and loved ones, takes it from the abstract and makes it very personal. Yet, given the divided world we live in, it is not clear that there is an obvious pathway that can lead us forward and help overcome the challenges that humanity faces.

What is the overall new future we want? Not from a narrow, siloed perspective, but holistically, including many interacting adjacencies and issues of long-term sustainability. Perhaps nothing is more critical than understanding what that final goal is, not in technical terms, but in easy-to-understand human terms.

For example, reducing atmospheric CO_2 levels to <300 ppm is not a humanly relatable goal. Most people do not want to generate any CO_2, they only want to live their lives – the CO_2 is an undesired outcome.

Most people do not want to generate CO_2 intentionally, they only want to live their normal lives. CO_2 is an undesired outcome.

Once a minimal set of goals is agreed upon, can we then set up metrics to assess that we are making progress toward these goals, and whether the proposed strategies can achieve scale (with all the glorious complexity that this invokes) in a meaningful timeframe and at acceptable cost? Can we define solution attributes and requirements, as well as a roadmap to get us from where we are to where we want to be? We hope this book will provide a fresh perspective on where we think we want to go as a society, and to discuss pathways that could get us there.

How Did We Get to Where We Are?

We live in a world today that has been shaped by over 6000 years of human ingenuity and innovation. Pursuit of fact-based science and technology has allowed us to unlock secrets of how the universe operates and to use this understanding to make our lives better (and sometimes worse!). Today, the changes, driven by science, technology, and innovation, are happening with unprecedented speed, and are driving impact that can be extreme in many ways; again, making our lives much better or much worse! Surely, science and innovation can guide us through the new upcoming challenges as well. The journey from scientific discovery to innovation to market adoption and finally impact at scale is at the heart of how we have achieved wave after wave of new innovations that have enhanced our lives and completely transformed our world – from an agrarian society to the modern world we live in today. But this could not have been achieved without generations of scientists, engineers, investors, entrepreneurs, and businesses (including their employees) working across the globe to solve tough problems, and to help their customers adopt these solutions. Finally, government, policy, and finance are major factors that significantly impact the success, scale, timeline, and equity of solutions that result in societal transformation.

If we just look at the history of how transformative ideas have moved from science to impact at scale, we typically see long time horizons, often stretching 50–100 years and more. If we look at automobiles, the first prototypes built in 1859 showed the way, but automobiles finally reached the masses and gave them the boon of personal mobility well after World War II, almost 100 years later. Ubiquitous access to electricity, even in the developed nations, took almost 70 years, and is still at best a work in progress in many parts of the world, more than 140 years later. Even for PV solar cells, first conceived and built in the 1950s, it took almost 70 years before they became competitive in the broader energy market. On the other hand, we see technologies like mobile phones and the internet, where service is now available globally to over 6 billion users over a much shorter period of 20–30 years. If it takes us 100 years to address the issue of carbon emissions and its adverse impact of our life, then we will have achieved little. Can this be done in 20 years? Do we understand the fundamental drivers that make some technologies move fast, while others take forever? Can we systematize the process, reduce risk, and assure positive outcomes – thus unlocking the flood gates of investment and adoption? Can we accelerate the process of scaling for positive impact, while avoiding adverse impacts, thus addressing the proverbial all-important Law of Unintended Consequences before it presents an impenetrable barrier in our path forward?

To understand what we need to do to get to a more desirable future, we first need to understand how and why we are where we are, as well as where we want to go. Maybe that will provide us with guidance on why things now need to be done significantly differently from what our past experiences suggest, and where new thinking may be needed. If there is uncertainty about how the future will evolve, do we freeze in our tracks and make no major bets, or are there "no regret" investments that we can make today that will support our journey along a number of different but likely scenarios as they evolve, which can collectively move us toward the primary goals that we have set? Are there simple fundamental principles underlying today's complex energy infrastructure that will provide clarity on what type of energy system is needed for the future? Do we know how to realize such a system, and where the gaps are, if any? Given the trillions of dollars invested in today's infrastructure, the tens of millions of people who work in related industries, and the billions of people who benefit from the existing energy system – it is clear that any major change has the potential to disrupt people's lives and economic well-being, at least in the near-term, and should, in the ideal world, be carefully thought through and managed. This will require alignment between all major stakeholders and a flexible and adaptable implementation plan focused on the fundamental questions.

Accelerating Change

As we look around us, we see accelerating change across a wide range of sectors related to energy. New twenty-first century technologies hold the promise of completely changing the manner in which our societal energy needs are met. Dozens of new technologies are carbon-neutral or carbon-negative and exhibit steep and sustained learning rates, promising lower costs and rapid scaling even while meeting global sustainability goals. These new technologies include solar photovoltaics (PV), batteries, electric transportation, blue and green hydrogen, CO_2 capture, water purification, direct air capture of greenhouse gases (GHGs), permanent sequestration of CO_2 and GHGs, new energy-efficient carbon-neutral methods for producing raw materials, chemicals and food, and replacement of fossil fuels with renewable carbon-neutral fuels. If we could wave a magic wand and move to a world that incorporated many of these technologies, we could certainly solve the anthropogenic carbon emissions problem – but could this be done in time, and what price (in economic and human terms) would we pay to get there? If the impact is so obvious and the outcome so desirable, then why is this discussion even happening? Don't we have all the resources of the government and private sector aligned to solve these problems and to roll out the new and improved tomorrow as quickly as possible?

In the USA we have the Department of Energy, the National Science Foundation, national labs, research universities, and major corporations, along with hundreds of similar governmental, nongovernmental, and private organizations across the globe, who are spending billions of dollars doing high-risk scientific research that will underpin the key technologies and businesses that can mitigate climate change, with many technologies now becoming visible and showing promise. We have governments, the World Bank, International Finance Corporation (IFC), major corporations, venture capitalists, and investors all focused on taking energy technologies to market and scaling these to address our climate challenges. Philanthropic funding from billionaires and from large investment funds is focused on commercializing technologies that will significantly reduce the level of CO_2 in the atmosphere in a 20 year span. In the USA, states such as California, along with many countries in Europe, are establishing policies that are forcing change – requiring electric vehicles and higher renewables on the grid. These policies have in turn triggered a tsunami of development with many new products becoming available to meet the new requirements.

We feel this is all wonderful and is very much needed. *Yet, we are worried by a nagging doubt that there are significant gaps.* A desire or wish is aspirational and does not constitute process, strategy, or an executable plan. Moving fast does not mean we are moving in the right direction, especially when what we do impacts others who are moving equally fast, but on independently determined trajectories and without much coordination. Availability of significant funding brings large numbers of recently minted "experts" and solution providers to the feeding trough, who all claim that theirs is the right solution. Loud voices proclaim every little success on the internet, social media, and the press, creating a cacophony that is difficult to pierce through. Whether it is solar, batteries, EVs, hydrogen, smart grids, microgrids, nuclear fusion, small modular reactors, carbon capture and sequestration, biofuels, or electric aviation – there is no shortage of researchers and companies proclaiming victory and taking issue with all other competitive solutions. This is normal with the rollout of new disruptive technologies, a lot of jostling for position as the evidence accumulates and winners and losers emerge. Unfortunately, this all takes time – which we do not have. Is there an alternate method for mitigating risk while accelerating the adoption and scaling of multiple competing solutions? After all, the total available market for new energy solutions is huge, and it is in the best interest of policy makers and governments to make sure that the end goal is achieved, and that scale is reached quickly and cost effectively.

Unique Opportunity Ahead

We believe that new and rapidly evolving twenty-first century technologies with steep learning rates, high modularity, and the ability to scale rapidly are offering us a unique opportunity to reimagine the world we live in. We are at the cusp of a fundamental transformation in our lives, and many things we take for granted will need to be revisited. A global economy built on the exploitation of scarce and limited natural resources that creates geopolitical tensions and exacerbates inequities should be out! What should be in is a society that is powered by abundant, affordable, resilient, and sustainable energy that uses universally available raw materials, where technology supports local manufacturing and wealth creation, and where sustainability is ensured by a circular economy. *We believe that deep decarbonization should be an outcome of doing things that are economically the right thing to do, and where minor tweaks to policy can initiate the virtuous cycle.* For the first time, there is an opportunity that these new technologies will be good for our wallet, and also for the planet! Done right, we feel there is a good chance that this can also be accomplished in a timely manner.

> *For the first time, there is an opportunity that these new technologies will be good for our wallet, and also for the planet!*

The foundation for all of this is the energy infrastructure, which powers everything and is the focus of this book. We feel that success not only requires development of new science and technology, but also a restructuring of the processes and mechanisms to accelerate the translation to economic value and impact at the societal level. The focus should be on de-risking all aspects of the journey from concept to impact at scale. This also involves directing leveraged resources (such as government funding) toward solutions that will address the global issues and will interoperate with each other to rapidly achieve scale. The approach should also consider issues of finance, policy, incentives, and the intersection with legacy systems, so that win-win solutions can be formulated where possible.

In this book, although we have generally tried to take a broad look at all energy issues, we have also focused on the electricity sector for a couple of reasons: first, we believe that electricity will play a critical role in the many possible energy futures that can evolve over the next 10–20 years. Secondly, we personally work at the intersection of the grid and some of the new emerging technologies, and we are seeing key concomitant challenges emerging that will impact our ability to reach scale in a timely manner. *Finally, because electricity is a mix of fast-physics, complex policy, and economics, an oversimplified view of the disruptive changes that are occurring will be unable to guide us as we move forward.* We need to look at the challenges holistically, thinking broad and deep at the same time, paying particular attention to things that are different at a fundamental level from our past. Managing this energy transition is critically important on many fronts – if we don't do it well, we will pay a heavy price for it, in terms of climate, carbon, cost, economic growth, and major disruption and chaos that can result if things are not done right. But, if there are so many questions, how will we know the right way forward?

This book starts with trying to understand how 6000 years of history have shaped how we think about energy and tries to explain why we are where we are. We feel that it is this interplay between science, technology, innovation, investment capital, and policy that needs to be understood. It helps us understand why we failed to anticipate the period from 2000 to 2020, and why that period is very special and positions us to do things that simply could not be done before. The rapid pace of technological and societal change that we are living with also calls into question the very nature of our institutions, our research universities and national labs, and the role that they and the government need to play as we move ahead. The very process of venture capital and forward-leaning investments, especially as it applies to high technology and innovation risk opportunities, deserves scrutiny. The role that regulators can play in shaping the energy infrastructure of the future also needs to be elaborated. As one can see, these are complex issues, but unless we take a broader, and at the same time deeper, view of the issues, we will not be able to formulate what needs to be done.

The rapid pace of change occurring all around us, in EVs, PV solar, wind, microreactors, hydrogen, and smart grids, suggests that we are approaching a tipping point, where our energy systems, and our lives, will be irreversibly changed sometime in the near future. What we don't know is if this change will move us toward a dystopian future, or one that is desirable and that we would want to leave as a legacy for our children and grandchildren. Time is short and the misalignment between economic needs and decarbonization poses challenges in terms of reconciliation. But there is hope. Today, we have the solutions that can give us both: economic well-being, as well as a sustainable future. If we understand the nature of these innovations and the process by which they can scale and have impact at a meaningful level, we may have a better understanding of what we must do next.

This book operates on a deep belief that everyone is right, in their own frame of reference, and that people generally put a higher priority on protecting their own interests in the near to mid-term, than on long-term issues that have a more distributed impact across broader society. This means that they are not averse to tackling the long-term issues, provided the near-term is not adversely impacted. It is up to us to forge a path forward that achieves both. We think this is possible today and can get to scale over the next 20 years. It requires that government play a critical role in guiding this transition forward, particularly by setting the broad societal goals and the guard rails that ensure they are met; helping accelerate rapid scaling of the right solutions; enforcing sustainable cradle-to-grave management to protect the environment; and preserving the rights of future generations. Scientists, technologists, entrepreneurs, investors, and leading companies are ready to take these new opportunities and turn them into reality. Once we have de-risked and validated technologies that have been proven in terms of value delivered, there is no limit to how efficiently and quickly they can scale – if they are given the freedom to operate within well specified and enforced constraints. This can be done, but there is only one chance to get it right!

But to understand what needs to be done, we first need to understand the history of energy and all the interconnected factors that have brought us here.

References

1. UN Sustainability Goals. https://sdgs.un.org/goals.
2. Adapted from HDI vs. Electricity Consumption Data (EIA, 2017; Jahan, 2016).
3. The Future of Electric Power in the United States. (2021). The National Academies Press. http://nap.edu/25968

Historical Perspective on Energy (4000 BCE–2000 CE) 2

Early Energy Sources and Uses

Energy consumption and human development are deeply interconnected, with more advanced countries generally showing higher levels of per capita energy consumption. Over millennia, the economic and military might of countries has been projected by their ability to control and access natural resources, and to transform these resources into valuable goods and commodities. Our modern society would cease to function if we did not have access to sufficient energy to meet our many needs. It is not surprising that our journey over the last 6000 years is deeply and inextricably intertwined with energy, and its availability or lack thereof. As our needs have grown, so has our dependence on energy, and our very world has been shaped and influenced by its context and story.

The first chapter in this story is our transition from nomad hunter-gatherers to an organized society. This transition depended on our ability to harness agriculture and to control fire – using it initially for cooking, light, and heat. The source of this energy was biomass material, such as wood or brush, or animal-derived fats and oils. The first productive kilns, dating back to 6000 BCE, were found in Iraq and were used to fire pottery and make bricks. By 4000 BCE, people had mastered the art of achieving temperatures high enough to melt tin and copper, allowing them to shape the metal into tools, trinkets, and weapons – ushering in the Bronze Age. The use of coal allowed even higher temperatures to be reached, allowing for extraction, and smelting of iron, leading to the Iron Age and establishing metallurgical principles that are still in use today.

While thermal energy in specific applications could achieve high intensities, the ability to harness and convert other forms of energy remained poor. The first use of wind energy was for sailing and dates back to around 4000 BCE. However, to achieve mechanical motion that was fully under control, energy was typically provided by human and animal power, and was used for pottery, irrigation, graining, farming, transportation, and the movement of materials. Irrigation could be

D. Divan, S. Sharma, *ENERGY 2040*,
https://doi.org/10.1007/978-3-031-49417-8_2

accomplished using human or animal powered buckets for lifting water from wells or rivers, or by diverting and managing river water with dams. Farming required ploughing, graining, and threshing, which needed linear, reciprocating, or rotary motion, which could be applied to till the field, shake the grain loose, and then finally crush the grain to make flour. Material movement was very important for trade, for things like mining and to build larger monuments, and was done over water using barges, or over land using human or animal power. Complex societies developed in Mesopotamia, Egypt, and India around 4500–3500 BCE, all based on their ability to harness energy, organize activities, river valley resources for agriculture, and to feed their people.

Concepts of simple machines, such as levers, pulleys, and inclined planes, attributed to Archimedes around 200 BCE, but probably in use in some form well before that, provided a mechanical advantage that could leverage available human and animal motive power. Coordinating and aligning the efforts of hundreds of people and animals also allowed our ancestors to do amazing things, such as building large monuments that have survived into modern times – such as the pyramids in Egypt starting around 2800 BCE, and Stonehenge around 2500 BCE. By 3500 BCE, the first wheels were being used, although it was much later – by 1500 BCE or so – that spoked wheels allowed bullock and horse-drawn carts to move heavier goods and to travel longer distances. This transition to wheeled transport over land helped to dramatically reduce the energy requirements for moving material (including people) from point to point, especially if the movement occurred over reasonably flat and smooth roads.

Overall, the primary sources of energy available to our ancestors were sun, wind, water, biomass, and fossil fuels. Sun and wind were low-grade resources, and water as an energy resource was very specific in terms of where and when it could be used. Biomass and fossil fuels provided a much more controllable and easily transportable energy resource, which could also provide light and heat energy with sufficient intensity to enable societies to increasingly organize into cities and kingdoms and achieve economic growth and to project military power. But thermal and biomass energy could not provide mechanical motion.

Our ability to realize mechanical motion remained linked to motive power that was supplied by humans and animals. In addition to being very limiting in terms of what this energy could achieve, it was also very poor in terms of overall efficiency. Humans and animals needed food to be able to work. This in turn required that appropriate biomass be grown for consumption by both humans and animals. In the case of humans eating meat, the energy cycle was even less efficient, as the biomass was consumed by the animal, only a part of which then provided food for human consumption. The biomass itself eventually derived its energy from the sun through photosynthesis – a process that is very inefficient (approx. 0.5% of incoming energy is converted to phytomass) [1]. As a result, arable land, water, and human and animal action needed for farming placed limits on the density of population centers that could be sustainably supported. Rather than the amount of phytomass that could be grown in the world, the major limiting factor was the inability of our ancestors to scale the amount of useful mechanical work that could be done through human or animal effort. As it turns out, the ability to scale new innovations is a critical parameter that would constrain what humans could achieve over thousands of years and would severely limit progress that could be made at a societal level.

As it turns out, the ability to scale new innovations is a critical parameter that would constrain what humans could achieve over thousands of years and would severely limit progress that could be made at a societal level.

Early Mechanical Energy Conversion (4000 BCE–1700 CE)

Conversion of latent or potential energy to mechanical motion that did not require sustained human (or animal) effort was recognized early on as a very important need. For instance, the earliest water wheels for graining, irrigation, and water supply date back to 4000 BCE. Over almost 2000 years, water wheels slowly increased in sophistication and were used in varied applications – including for iron smelting, grain threshing, water lifting, and mining. The simple machines that were already known provided further mechanical advantage that could leverage the perpetual motive power that was now available. However, the use of water wheels was constrained by its specific location and the availability of running water or water with sufficient head – as such it could not be universally applied. The lure of using wind for applications other than sailing was also strong, with attempts to harness it going back to around 2000 BCE. However, successful windmills would only be realized at a much later time.

War has had a major role in shaping human technologies and developments. Using mechanical advantage to hurl arrows and missiles at the enemy required the development of increasingly sophisticated bows, slings, catapults, and trebuchets. All of these were in the category of simple machines, where human or animal energy was accumulated and rapidly released as kinetic energy in the missile. This remained the status quo until around 900 CE, when gunpowder was first formulated in China by mixing together incendiary and oxygen providing chemicals that were already in use. For the first time, humans had the ability to control the time and location at which a rapid conversion from chemical to kinetic energy could be initiated. This eventually allowed the development of guns and cannons that could hurl large projectiles hundreds of meters, causing immense damage to armies, stone walls, and fortifications from a safe and long distance away. The use of gunpowder spread rapidly around the globe and continued to see refinements that made guns lighter and more maneuverable, while also making them more deadly at the same time. While this clearly accomplished the objective of unlocking high levels of kinetic energy, the release of energy was explosive and not well controlled. As a result, this did not lead to improvements in the efficacy of mechanical energy conversion, which had to wait until the invention of the steam engine more than 700 years later.

The European Renaissance, lasting from around 1300 to 1700 CE, signaled a transition from the ancient world to the modern world that we live in today.

Developments in art, philosophy, science, and technical advances such as the printing press, established many of the elements of today's culture. This was a time of discovery in Europe – Marco Polo to China in 1290, Columbus to the Americas in 1492, and Vasco da Gama to India in 1498 – showed Europeans the world was accessible. Sailing ships from Europe had become more reliable and provided access to the riches of the East, and the potential to occupy vast new lands for natural, mineral, and human resources. This eventually led to colonization by the European nations, allowing them to plunder wealth and resources from colonized territories to grow their own economies.

At the same time, the 1500s saw an increased demand for coal to provide heat for urban areas, and for finished goods such as textiles and woolen cloth from England. This demand triggered a need to improve the efficiency in producing and delivering these products. The need to improve mining and manufacturing drove the search for an engine that did not depend on, and was not limited by, human or animal power. Today, we consider science, technology, invention, and innovation as inextricably linked with each other. That was not the case in the 1500s and 1600s. In that era, science was more in the realm of philosophy and knowledge, and very few institutions and universities had formal programs focused on scientific discovery. Instead, the early European universities built on the classical teachings of Aristotle, Plato, and the scriptures. Over time, they branched out into the natural sciences, mathematics, astronomy (or rather astrology), theology, law, and the arts. Technology for practical use was developed by individuals who were self-funded or funded by benefactors and involved tinkering and experimentation with what was experiential and observable.

Despite this, many scientists made major advances that can be attributed to this period – including Copernicus who was the first to propose a heliocentric view of the world; Galileo who invented the telescope and was a prolific writer on diverse science issues; and Newton who first propounded the concepts around calculus and the three laws of motion. However, none of these would address the immediate need for a machine that could provide the energy conversion services that society wanted. At around the time that Newton laid the foundations for calculus in 1666 and published *Principia Mathematica* in 1687, Thomas Savery was trying to harness the power in steam and convert it into mechanical motion. He patented his steam machine for converting heat energy into mechanical work in 1698, specifically targeting mills and mines, with an objective of providing them with controlled mechanical motion for pumping and to operate looms. The power contained in steam was known in antiquity (the Greeks conceptualized the aeolipile, a steam operated machine, around 0 CE). Other examples from China, India, and the Middle East can also be found. Starting in the 1600s, much scientific work was also done around the principles of pressure and vacuum by many including Torricelli, Boyle, and Papin. However, Savery was the first to patent and build a practical (though not very good) engine specifically for an industrial application. This really marked the beginning of a new era in energy conversion and use, and completely changed our societal trajectory forever.

Coal and Mechanization (1700–1880 AD)

In many ways, a new era of harnessing increasingly nonintuitive science and technology into useful inventions and commercial value could be said to have started around 1700. Savery's steam engine showed that it was possible to use heat from wood or coal, and to automatically convert that into continuous mechanical motion that could power useful revenue generating applications. However, its poor efficiency (<1%) limited the use only to specific applications and the device could not be broadly applied. On the other hand, a brand-new method for achieving this type of energy conversion was now validated. It was now possible for a multitude of new inventors to tinker with the limitations of the proposed Savery engine, and investors/patrons to bet on the success of their particular version. This technology showed it was now possible to free humanity from the limitations of human and animal power and established a trend that appears to be irreversible. In the language of contemporary research and development, a "minimum viable prototype" (MVP) had been built where the innovation risk was eliminated – thus unleashing a continuous wave of innovations (new "Use Cases") over the next 150 years and more. Although it took hundreds of years for the benefits of this mechanization to flow to all strata of society and to all nations, the Savery steam engine showed us the possibilities.

In the language of contemporary research and development, the Thomas Savery's steam machine, patented in 1698, was a "minimum viable prototype" (MVP) where the innovation risk was eliminated – thus unleashing a continuous wave of innovations over the next 150 years and more.

A process for taking innovation to market was gradually being developed, not by design but through coincidences and individual proclivity. From 1700 until around 1780, the steam engine continued through a series of improvements, driven by inventors including Newcomen, who was influenced by the work of Boyle and Papin, and developed the first engine that used pistons and cylinders for higher efficiency and worked with steam at atmospheric pressure for improved safety. However, he had to partner with Savery (due to patent infringement issues) to produce the more efficient engine. This engine could finally pump water from deep mines also (which the Savery engine could not do), resulting in more than 100 Newcomen engines being installed by 1735.

Perhaps the biggest leap in steam engine technology came in 1776 when James Watt developed an engine with a separate steam condenser that allowed cylinder temperature to be maintained and dramatically improved engine efficiency. By 1781, he had also developed a mechanism to convert the reciprocating motion of his steam engine into rotational movement, allowing it to address a much wider range of applications. As is always the case, the path from concept to commercially viable deployment was very rocky. Manufacturing precision cylinders and pistons in that era was problematic and drove some of his financial partners into bankruptcy. By

1800, the company of Boulton and Watt was well established and was successful in producing about 500 steam engines. Innovations continued, including the use of higher-pressure steam, a double acting engine and improved valve regulation and actuation. These innovations allowed substantial increase in power produced, efficiency, and power density (reduced size) – metrics that continue to be very important even today [2].

These advances also allowed steam engines to jump from land-locked industrial applications to mobile applications – in particular, railroads and steamships. The first steam locomotive was tested on a track by Robert Trevithick in England in 1802. By 1804, the first steam locomotive was in commercial use hauling iron ore from a mine and moving passengers shortly thereafter. The first commercial railway in England was operating by 1825 and had expanded to over 13,500 miles of track by 1870. The first commercial railway in the USA was operating by 1827, and rapidly expanded to span the continent by 1869. Growth of the steam powered railroad was global, with the first industrial railway operating in India by 1832 and the first passenger line in 1853, and in China by 1876.

As land-based transportation was being enabled by steam, so was movement on water. The first paddleboat powered by steam was operating in the USA in 1807, with the first transatlantic crossing by 1819. Voyage times decreased dramatically when compared with sailing ships and became more predictable. Passenger travel became more comfortable and passenger traffic and migration increased. The invention of the screw propeller allowed the building of larger iron-hulled oceangoing ships that did not require sails. The first such ship sailed in 1847, with rapid growth that soon saw the end of the era of sailing ships. The reduced dependence of steamships on wind patterns made shipping and travel more predictable and opened the world to trade and colonization.

The impact of these technology developments defined the world as we know it today. Britain's success in mechanization led to their emergence as a global power, driving colonization for access to raw materials for their factories and markets for the produced goods. This led to increased urbanization and slowly improved the livelihood of their people. The mechanization wave spread through developed nations, including a newly formed USA, driving economic growth. The mechanization revolution was initially powered by wood as a fuel, but quickly switched to coal because it was a more efficacious resource. Coal mining became a big business, and soot emitting smokestacks became a part of every major city's skyline – which the polluters and governments considered a small price to pay for prosperity! The ready availability of coal as an easily transportable source of energy, and steam as a new method for mechanical conversion, met a human need and triggered inventions across a wide range of sectors including manufacturing, refrigeration, farming, and metallurgy.

Why this became possible is best seen through a technology lens that compares key metrics for steam engines. For instance, the power that could be produced by a steam engine increased from around 2 kW for the Savery engine in 1700 to almost 100 kW for the Boulton and Watt engines around 1800. In another 80 years, this power had further increased to around 2 MW – an astounding 1000× increase in power in less than 200 years. At the same time the overall efficiency of energy conversion increased from around 0.5% in 1700 to 2.5% in 1800 and to as much as 20% by 1880 – a 40× improvement. This was accompanied by a dramatic increase in power density as well – from 1000 kg/kW in 1700 to around 60 kg/kW by the end of the nineteenth century – another 16× improvement! This also allowed the development of a wide range of machines of different sizes and ratings, each optimized for a specific mission. It now became possible to do things across a wide range of applications that simply could not happen with human and animal power.

Looking at the above discussions and on historical references to the Industrial Revolution, it would be reasonable for someone to think that the availability of the first steam engine marked the start of a new mechanized era. That was not necessarily the case. In the first 100 years, very few applications could afford to use the new technology, and the machines were specialized and had to be custom built by hand (remember that there was no mechanization or mass manufacturing!). The vast majority of people continued on with their normal lives, using people, oxen, and horses to perform the needed mechanical tasks. It is important to note that despite the appearance of the Savery engine in 1700, the slow and almost random process of innovation delayed the full impact of steam for 150 years [2].

But the overall impact on humanity was nothing short of transformative. In a span of 200 years, human ingenuity had transformed the world, accomplishing what the previous millennia had not been able to. This was the first major societal transformation to occur in over 6000 years – from Agrarian to Mechanized. It is what enabled the Industrial Revolution [2].

The higher profitability enabled by mechanization also increased the interest for the inventors and their benefactors to keep developing better and better solutions. The virtuous cycle was launched, and innovation could not be stopped. Without an efficient compact source of mechanical energy that could run factories, locomotives, and steamships, there would have been no Industrial Revolution, no colonization, and no mass migration. It is a testament to human ingenuity, invention, innovation, and the ability to overcome impossible challenges and to ultimately prevail – the world-transforming outcome of a collaboration between inventors, entrepreneurs, financiers and ultimately the people and the governments.

Even as we celebrate the accomplishments, it is also important to understand some of the issues and challenges that this rapid transformation created. What enabled success was the extremely high power and energy density of the primary energy resource – coal, and the ability to convert that energy into controlled mechanical motion in a device with an increasingly compact and dense form factor. The benefits to the nation and the economy were so significant that a blind eye was turned to issues such as increasing air pollution and smog, which contributed to health issues. Urban congestion, along with poor water, sanitation, and waste management, contributed to poor health and epidemics. However, scientific and medical knowledge was in its infancy and the cause and cure of diseases was a mystery. Residual waste from burning coal for steam contained high levels of poisonous metals that could affect drinking water. Even if these many issues could be seen and possibly understood, there was no easy way to track what waste was emitted and by whom. System-level interactions were poorly understood, if at all. The financiers, entrepreneurs, and governments were all interested in maintaining economic growth and improved prosperity, even if the rate at which it trickled down to the general population remained very slow, and the gains were very unevenly distributed. This does not even count the impact on those regions that were colonized or supplied slaves and bonded labor to meet the growth objectives of the "victorious" industrial nations.

As the number of factories using steam engines for stationary and mobile applications increased, there was a corresponding steep increase in overall energy use, and an increased concentration of wealth and resources in the hands of the people who were benefiting from the increased productivity – which widened the disparity between the haves and have-nots. It also created a growing need for raw materials and resources to support manufacturing, and to develop markets for the higher volume of goods produced. *That this also resulted in colonization, slavery, exploitation, and pollution was the pact we made with the devil, a pact for which we are all still paying a price!*

At the rates of consumption imaginable, it seemed that coal supply could extend forever, and did not raise any questions. The process by which coal had been formed eons ago was not understood at that time. The question of carbon dioxide as being a greenhouse gas was not understood – the mechanisms by which our planet maintained thermal equilibrium were not understood. The world felt like an infinite resource, which would always remain the same, and it was unimaginable (except for a few people like Malthus in 1798) that resources could become constrained, or that human activity could in any way impact the world we live in. It was however becoming clear that human activity was changing the environment, at least in urban centers, and not always for the better. At such a time, and in such a world, it is not surprising that the positive aspirational objectives far outweighed the negative cautionary issues, and that unfettered growth was pursued. Governments allowed industry to move forward at full speed, agreeing implicitly or explicitly to socialize all costs coming from this rapid industrialization and growth, providing them with a competitive advantage and setting the stage for future situations where industry would demand that such costs be socialized once again.

Science, Technology, and Innovation (1700–1880 CE)

This period bookending the Industrial Revolution also saw a dramatic change in the way we approached scientific knowledge and discovery. As discussed earlier, the traditional role of European universities, such as in Bologna, Padova, Uppsala, and Oxford, was to educate people in the classics, mathematics, and divinity, and to follow a method of memorization and regurgitation. This slowly began to change to a more holistic approach based on a freedom to question and to seek knowledge. Humboldt helped to start the University of Berlin in 1810 based on a faculty-led approach that included research and studies. Study of the sciences gradually increased – it was not until 1833 that the word "scientist" was itself coined (thanks to Whewell, a distinguished Professor at Cambridge). In 1876, Johns Hopkins opened as the first US research university, with many more following in the years ahead. There was an increasing understanding that science provided the underpinnings for technology, and that scientific knowledge and discovery was key to the innovations and technologies that could change the world. However, this did not mean that all universities were focused on scientific research. Even farther away was the idea of translating that research to value for society. That was the job of the company or the innovator/entrepreneur, not the scientist or faculty researcher. Application-oriented research was not considered pure and academic in nature, and was often frowned upon in universities – a trend that continues to this day.

Science and the Steam Engine

The transformation from an agrarian to a mechanized society is an amazing achievement, and it is tempting to think that systematic scientific discovery and technology development played a significant role in this process. After all, this resulted in a major societal transformation – and an upending of everything we knew or did before that. That was definitely not the case! As technology novices (in every field except possibly in our own area of expertise), we tend to have a simplified view of how technology development occurs and feel that experts are very knowledgeable, and that scientific knowledge and systematic processes guide the development of solutions. That may not always be the case!

For example, the design and operation of steam engines is covered by the laws of thermodynamics. The Carnot cycle, which defines the best efficiency that a thermal engine can deliver was proposed by Said Carnot in 1824, almost 124 years after Savery built his first engine. The most important and fundamental principles – the first and second Law of Thermodynamics were developed in 1860 by Clausius and Thompson, while the third law was only formulated in the early 1900s. This suggests that the development of the steam engine occurred without a sound theoretical basis or a quantitative understanding of the design trade-offs, but was supported more by observations, empirical data, and intuition developed over years of experimentation and validation in the field. The gap in theoretical knowledge was filled by hundreds of inventors and entrepreneurs, all tweaking the current designs through "trial and error" to solve specific issues that were limiting the applicability in a target market segment, with patent protection and market forces ensuring the survival of the fittest (not always the best) solution. This was a rather inefficient process, in terms of time, human and monetary resources, but it was the only one that we had at that time, and it has shaped our thinking on how the innovation and translation process works. This also raises the question: If scientific research and technological development could have been better aligned, would that have significantly accelerated the energy transformation?

The period from the 1700s to 1880 also saw tremendous achievements in science, many of which at that time probably did not appear to be significant or transformative for the field of energy, certainly not for society at large. Without question, luminaries like Newton and Galileo bookended the early part of this period with key scientific principles in statics, mechanics, gravitation, and astronomy. Bernoulli and

Euler laid the foundations for the principles of fluid dynamics between 1738 and 1752, analyzing flows of incompressible liquids and gases that were to become the foundation for how airplanes generate lift and fly – 150 years before this would actually happen. The principles of refrigeration, such as the extraction of latent heat from the environment as a liquid evaporates, were developed starting in 1755 by many researchers in universities. Following decades of experimentation by inventors and entrepreneurs, the first commercial deployment of refrigeration and ice-makers finally happened in 1880, followed by rapid expansion to feed the demand. This also needed a compact source of energy.

There were several other major research thrusts ongoing during this period that would go on to have tremendous impact. The first was the need for an engine that was more compact than a coal-based steam engine, especially one that could enable personal mobility. In the seventeenth century, experiments were conducted using gunpowder and controlled explosions to drive a machine that could pump water. Experiments in the eighteenth century continued and by around 1800 the first experimental two-stroke internal combustion engine had been developed. As with the slow development of the steam engine, hundreds of innovations were built on this foundation, evolving eventually into the modern four-stroke internal combustion engine first demonstrated by Otto in 1876. The commercial success of the internal combustion engine was enabled by a parallel thread of activity. Petroleum coming from the ground was known in antiquity, but wide use had not been found for this material. By the mid-1800s, it was shown that kerosene could be extracted from petroleum, allowing for a low-cost and readily available replacement for whale and vegetable oils, which were widely used for lighting at that time, but offered poor light. This opened up the possibility of cost-effective petroleum-based fuel distillates that could be used as a more accessible fuel that could be available at scale. This convergence of two distinct and different threads enabled the great mobility revolution that started around 1880 and created the oil and gas infrastructure that has defined so much of the twentieth century.

Scientists were also getting more interested in the area of electricity and magnetism. Although electricity and magnetism were known in antiquity, it was not until Volta invented a practical battery in 1791, one that could continuously provide DC current, that experimentation could reproducibly be done. Oersted in 1819 showed that flowing currents caused magnetic fields, while Ampere in 1820 quantified the force and evolved the principles of closed electrical circuits. Perhaps the most important early contribution was from Michael Faraday, who in 1850 demonstrated the principles of electromagnetic induction and the ability for currents in a magnetic field to exert force, and for motion of a wire in a magnetic field to generate electricity. Maxwell in 1864 published the equations that tied together electricity, magnetism, and light, suggested the existence of radio frequency waves, and provided the theoretical foundation for modern day electrical engineering.

Electricity as an energy source made tremendous progress from 1850 to 1880, driven by the availability of batteries, and the ability to build motors and generators based on the principles of electromagnetic induction. These developments would eventually lead to the development of the electricity system and power grid and are covered in a later section. Another related development was the need for more efficient turbines to convert water flow into mechanical motion. Francis developed a low-head turbine in 1848, while Pelton developed the Pelton wheel for high-head conditions. These would be critical for electricity generation at scale in the years ahead.

The work of these early scientists laid the foundation for hundreds of other scientists and inventors to build on these basic concepts and led to the evolution of the electricity sector, but as was typical of the era, science and technology were on independent tracks. Today we know that electric currents are caused by the motion of electrons, but the electron as a fundamental particle was not even discovered until 1897. At that time, however, much of the quantitative analysis was based on observations and did not come from a detailed understanding of the quantum physical phenomenon that underly electricity and magnetism. Instead, inventors and entrepreneurs, along with their financier friends, were trying to solve the practical problems that society faced, and they expected to make money by solving such problems. They did not wait for the science to evolve, but rather used experiments and intuition to develop practical solutions. Much of what they worked on was physically tangible and observable with the tools of the day. But science was catching up fast and would eventually open up brand new fields of exploration that would lead to new solutions – many of which to someone from the mid-1800s would have seemed like magic.

Changing Energy Paradigm (1880–1945 CE)

By 1880, the world had been transformed from an Agrarian society to a Mechanized society – the first major energy revolution. Along with that came urbanization, public transportation, ability to manufacture goods at scale, and to ship them anywhere – thus creating a global market. Incremental technology improvements would continue for many decades, but many limitations were becoming visible. These outlined the needs and created the market pull for the next generation of technologies.

Even as steam engines changed society by improving productivity and mobility, it was clear that they could not meet all our requirements. Steam had to be generated at point of use, could not be scaled down, was inflexible, needed skilled operators, and generated noise, waste, and smoke. There was a need for a more flexible and cleaner source of energy that could be used to fulfill a wide variety of end-use needs, ranging from residential to large industrial. There was also a need for a more flexible mode of transport that could be scaled down to the level of personal mobility and could operate without need for an elaborate rail infrastructure, making last mile access viable. The years leading up to 1880 had seen a flood of innovation around the ideas of internal combustion engines and electricity. These two paradigms gained ascendancy and ushered in a new era in energy.

Petroleum and Mobility: Point of Use Power

As steam engines and railroads took off, the need for last mile delivery and personal mobility that could operate without fixed steel tracks became pressing. Horse carriages were the dominant solution at that time, and as traffic increased, driven by increased prosperity and goods movement, the biggest concern for cities like London and New York was that they would drown in a deluge of horse manure! The earliest attempts at building cars started with Cugnot in 1796, who demonstrated a steam powered three-wheeler. Developments in batteries and motors in the mid-1800s led to several manufacturers offering battery powered electric vehicles in the period from 1880 to 1900, peaking at about 30,000 electric vehicles by the turn of the century. However, they would not survive the emergence of the internal combustion engine car that was also being developed in parallel.

The demonstration of a practical four stroke internal combustion engine by Otto in 1876, and its integration with key concepts such as the ignition coil, spark plug, carburetor, crankshaft, and distributor, enabled steady progression toward a commercially viable motor vehicle. At that time, even with a more limited driving range, electric cars were still preferred because they were easier to use, less noisy, and did not require a hand-crank to start the vehicle. As these issues were overcome, and the internal combustion engine-based automobile became more reliable and easier to use, it completely displaced the early electric vehicles.

As we saw with the steam engine, the slow pace of innovation and experimentation accelerated rapidly once certain viability risks had been addressed. Likewise, once innovation risk had been addressed for the basic concept of the internal combustion engine, hundreds of inventors, entrepreneurs and financiers jumped into the virtual "free-for-all," simultaneously improving on engine technology, and developing other support functions such as: car chassis, shock absorbers, gear boxes, differentials, pneumatic tires and the electric starter. Other related issues that were also addressed included manufacturing process, reliability, style, comfort, fuel availability and filling stations, improved roads, servicing and repair, and support. To get an idea of the scale of automobile manufacturing that was triggered, it was 1895 when the first car company was established in the USA – the Duryea. *Since then, the USA has had 1900 different car manufacturing companies, and 3000 different makes of cars that have been sold.* Early cars were intended for the wealthy but moved quickly to cover the broader population. Ford's Model-T, introduced in 1908 to an eagerly waiting customer base at a price of $850 (which dropped to $290 by 1924), exemplified the focus on creating a car for the mass market.

The development in 1898 by Rudolf Diesel of a first compressed-charge compressed-ignition engine, with its much higher efficiency, opened the possibility of extending the internal combustion engine to higher power levels. It was being used for ships by 1903, submarines by 1904, trucks by 1908, and to power locomotives by 1912. By 1925, the use of steam engines for ships had peaked. Similarly, diesels dominated all heavy use applications by the mid-twentieth century, with examples today of diesel engines generating as much as 100,000 hp for ship propulsion.

Another major application for petroleum that emerged was in aviation. In 1903, the Wright Flyer demonstrated the first heavier than air machine with sustained flight, powered by a 12 hp internal combustion engine. The high-power density of the internal combustion engine and the high-energy density of liquid fuels made aviation viable. The first commercial flight carrying passengers was in 1914 in the USA and 1919 in Europe. The first use of aircraft in war was in 1911 in the Italian-Turk War, and in World War I in 1915 onward for reconnaissance, aerial warfare, and bombing. The first trans-Atlantic flight in 1919, followed by Lindbergh's solo flight in 1927, paved the way for global passenger travel. The 1930s saw rapid growth of the global commercial airline industry, with the development of aircraft such as the DC-3 and Boeing 247. Aircraft played an outsize role in determining the outcome of World War II and resulted in the development of many new technologies including the first jet engines in 1941. Jet engines eventually helped to shrink the world through the development of air travel that was also affordable for the masses.

The availability of liquid petroleum-based fuels completely changed societal mobility, shrinking the globe, enabling migration at an unprecedented scale, and helping to create a global economy. It also served to mobilize armies on ground, at sea, and in the air, using this mobility to project strength. As an example, a major factor in the Allied victory in World War II was the ability to limit German access to fuel for their aircraft, tanks, ships, and motorized vehicles. Germany tried to offset dwindling supplies of petroleum by developing the technology to make synthetic fuels from coal, the plants for which were then also targeted by the Allied forces. By the end of World War II, the global transition to petroleum-based mobility had been completed, and a new race was on – to supply the global economy with the oil it would need to meet growth and nation rebuilding needs.

Petroleum consumption in the USA increased from less than 0.1 Quads in 1880 (petroleum and gas combined) to 10.11 Quads by 1945, most of which was used for mobility (this increased to around 36.7 Quads by 2019). By way of comparison, the total energy consumed in the USA in 1880 was around 5 Quads (2.9 Q for coal and 2.1 Q for wood), increasing to 17.3 Q by 1945 (16 Q for coal and 1.3 Q for wood) (Fig. 2.1).

By 1945, the USA was a net exporter of oil, supporting the recovery in Europe and Japan. However, its own consumption was rapidly increasing, as was consumption around the world. New economic superpowers who derived their strength from oil would soon have an outsized impact on the global economy.

Electrification: Power at a Distance (1880–1945)

Petroleum cleared the path for compact energy sources that enabled mobility and point of use applications. Internal combustion engines could also replace steam and provide local motive power for processes and industry. However, the need to have a power plant in every location that energy was needed, including homes, businesses, and industry, required complex machinery and skilled staff at each location (remember this was before microprocessors and automation). This was very expensive and

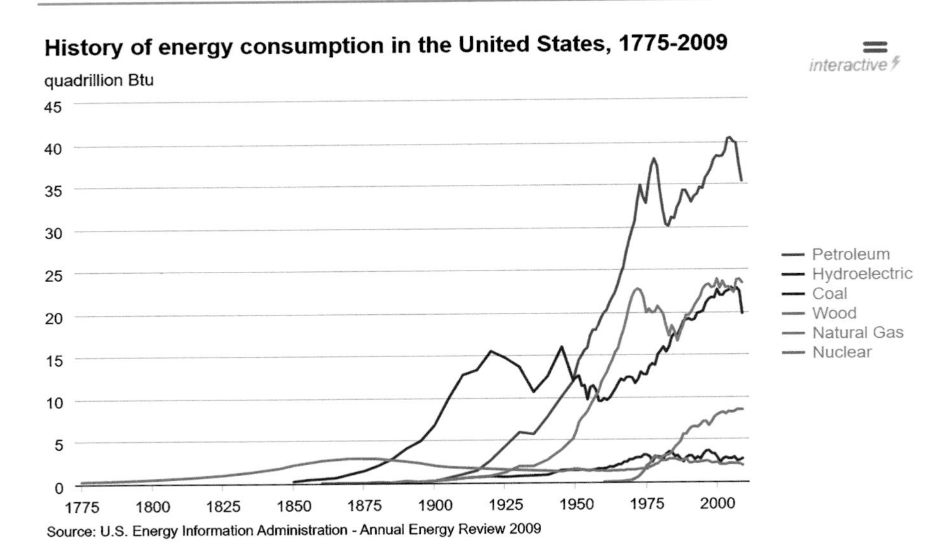

Fig. 2.1 History of energy consumption in the USA [3]

simply not practical. Also, thermodynamics limited the efficiency of compact internal combustion engines – an acceptable compromise when mobility and generation at point of use were critical. What was needed was a source of energy that was economical and efficient to generate and transport, was readily accessible at each end-use location, could be used by relatively unskilled operators and lay public, and which could power a wide variety of applications – all at a long distance from where it was generated. This was electricity!

By 1880, batteries had matured considerably, driven by a constant stream of innovation and tinkering. Early applications, such as the telegraph, were not energy intensive, and could be powered by these new emerging batteries. However, higher power for long durations proved to be more challenging. The development of the lead acid battery by Plante in 1859 showed a battery that could be recharged several times, leading to Faure's design of a more manufacturable and reliable lead-acid battery in 1881. At the same time, Leclanche's work in 1866 led to improvements by Gassner in 1886 and to the first "dry cell" battery, which did not involve liquids and noxious gas emissions. These two capabilities further spurred the rapid improvement of battery technologies that made sustained DC power available to support a wide range of applications. For electricity to be really useful over a wide range of applications, it was also important that it could be converted to mechanical energy and vice versa. Faraday's work in 1831 triggered tremendous innovation around the world. Von Jacobi developed the first practical electric motor in 1835 and used it to power a boat. In 1837, the Davenports obtained the first US patent on electric motors. However, the short life and high cost of batteries prevented many of these ideas from reaching commercial success.

Like we saw with the steam engine, the demonstration of a minimum viable prototype and reduction of innovation risk was followed by a broad wave of technology development and tinkering. In 1838, Stimpson developed the segmented commutator followed by Pacinotti in 1864 who developed the ring armature to make DC motors practical. Siemens in 1867 showed the ability to reverse power flow, making the same machine act as a motor delivering mechanical energy, or a generator that received mechanical energy and transformed it to electricity – a critical realization. The first practical integration of many of these technologies was done by Edison, a prolific inventor and entrepreneur, in 1882 with the Pearl Street Station generating plant, integrating a steam turbine with a DC machine to produce power at a "standard" voltage of 110 volts DC to power electric public lighting in New York. This rapidly drove cities around the world to replace gas lighting with electricity and started a wave of electrification.

Edison also understood the importance of generating power centrally and then delivering it to homes and businesses to use in varied applications. He saw early on that to succeed he would need to develop and provide all elements of the system – including generation, power delivery, and end use. Nothing was available as this was a new and disruptive industry. Even as DC was rapidly being deployed, its limitations were becoming clear. DC electricity generation, involving coal and steam, had to be within a short distance of where it was used because of voltage drops along the cables. Higher voltages could have allowed a wider distribution area but was deemed too dangerous. The solution was to build generation plants within city blocks, running heavy cable in underground trenches to houses, streets, and businesses where the electricity would be used. It was clearly preferrable that generation be located in a distant place where energy sources were available or were economical, and the power be brought in – something that the DC power of the era could not achieve.

It was recognized early on that rotating an electric coil in a magnetic field generated an alternating current (AC), which was then converted to DC using mechanical commutators (which had reliability, life, and safety issues). Much work was done to see if the AC current could be directly used in power generation and utilization. Ferraris in 1885 developed the concept of a polyphase induction machine. Working independently, Tesla, in 1888, patented several AC machines, including induction machines and a synchronous machine with separate DC rotor excitation – which was the precursor to today's large fleet of AC generators, technology which was quickly licensed by Westinghouse. Such an AC system could do all that the DC system could do, but also shared many of the same limitations. However, this was about to change.

Building on principles established by Faraday and early work on induction coils, the ZBD transformer was developed in Hungary in 1884 that showed that AC voltages could be stepped up and down. Since power is the product of voltage and current, higher voltages meant the same power level could be produced with lower currents. This implied that AC electrical power could be transmitted over longer distances without significant loss of deliverable power – something that was not possible with low voltage DC systems. Westinghouse licensed the patents for the

ZBD transformer, and then worked with Stanley to demonstrate a commercially viable transformer in 1887 using a closed magnetic core and stacked laminations – a design concept that persists to this day. The transformer gave AC systems the ability to step up or step down the voltage at will. This finally enabled the use of distant steam or hydropower generation, from where the voltage was stepped up so it could be sent over long distances, and where it could be stepped down again to a safe level before it was delivered for use by lay consumers. Such a centralized generation system could serve thousands, and eventually millions of customers.

The period from 1880 to 1900 saw the rapid growth and change that is characteristic of new disruptive and transformational technologies. Hundreds of technologists were tinkering around trying to solve problems that were inhibiting commercial success and helped to serve a huge need that had been identified and for which tremendous market pull existed. Along with steam turbines, which were now available, the work by Francis in 1849 had led to the successful development of hydropower turbines that could supply enough power for entire cities. To be able to deliver power to distant load centers, experiments were conducted for transmission of DC power with a first commercial deployment in 1882. Steinmetz developed the underlying mathematics for three phase systems, showing that it would be the ideal configuration for high power AC systems. In 1891, the first 3 phase AC power system showed power transmission over 175 km in Germany. In 1893, the first commercial 3 phase system was deployed in Redlands, CA. The first truly commercial AC power plant was built at Niagara by GE to supply power over a distance of 25 miles to Buffalo. By 1900, deployment of AC systems exploded over the entire world. Electrical power had come to the mainstream.

However, unlike internal combustion engines, where every automobile is generally very different from other vehicles (except in the use of "standard fuels"), in electricity, all customers needed to interface with a common power delivery system. Without any coordination, individual entrepreneurs and companies were optimizing for their own success. This resulted in wide disparity in specifications and parameters of the system. For instance, in the early days, operating frequency ranged from 25 Hertz to 133 Hertz, and voltages from 100/500 volts at point of use, and 1000 to 14,000 volts for high voltage transmission. Each company wanted their own technology to succeed, creating terrible fragmentation. The lack of interoperability made it very challenging to develop end-use applications that could be used by a broader market and prevented the market from scaling to its potential. Of perhaps even more importance was the inability to store electricity. This required instantaneous balancing of generation with connected load – something that was counterintuitive for nontechnical people more accustomed to dealing with commodities that could be stored and traded. Inability to store electricity was less of a problem for small local generators that individually supplied their loads but was a big issue when multi-generator geo-dispersed systems were considered. All these factors presented what seemed to be insurmountable problems for this nascent industry that was trying to grow.

Further, given the infancy of the industry, there were no standards to guide the engineers. Many of the basic questions were not even understood. At the same time, companies were concerned about maintaining competitive advantage and proprietary intellectual property (IP) that would allow them to commercially succeed. It was in everybody's best interest that a common framework be established that would allow the overall market to grow, allowing each company to then compete for a share of this massive new market. Further, an inability to reach a common framework and set of operating rules, could result in a much smaller market, slower growth, poorer safety, and much confusion among an unskilled user base. As we will see in later sections, such issues are highly relevant to our current energy crisis in the twenty-first century as well!

The primary coordination issue was not about trade and policy (which was challenging enough), but about controlling a physical system with many devices geographically dispersed over hundreds of miles, that were all very large (weighing hundreds of tons each), moving at high speed and were electrically interconnected, and as a result also "mechanically" coupled through the electrical system (even a small deviation could create tremendous stress on the components as the system tried to get back into equilibrium). In this geo-dispersed system, things had to be coordinated with millisecond precision because electricity could not be stored, and absolute safety had to be guaranteed under normal and abnormal conditions often in the presence of lethal voltages and with unskilled end users. The electricity system would not scale if these challenges could not be solved. We should remember that this was a world before microprocessors, communications, or fast digital controls – so there was minimal automation and only very primitive control capability.

We had to develop generators with intrinsic properties that allowed them to interoperate and collaborate with the many other devices from different manufacturers at different locations that were all interconnected. They had to function so that they shared power and energy across a wide range of operating conditions, from light loading to heavy loading. We needed to grow the system, attaching more generation as needed, repairing devices when needed, and ensuring that faults at some location on the system did not bring everything down. We needed a system that provided safe, affordable, and reliable power, whenever and wherever we needed it! We take this for granted today, but it was really our ability to solve these very challenging technical issues using a cohesive set of rules that enabled an electricity grid that today covers every part of the globe and has been called the largest machine built by humankind and its most outstanding achievement and was named as the biggest engineering accomplishment of the twentieth century by the US National Academy of Engineering.

How could this monumental achievement of "electricity at a distance" happen in a world with fast evolving technologies, poor understanding of a system that had not yet been built, lack of a regulatory and policy framework, and a constant tension between enabling a large market and giving away the proprietary IP that could give a company a competitive edge in the market? Our apologies to the gentle reader for this rather deep dive that we take below into how the grid of the early to mid-1900s worked and evolved. It will hopefully inform future discussions on why the grid does what it does today and how the future grid is continuing to evolve. What is perhaps most remarkable is that the basic principles of how the grid operates today, more than 100 years later, have essentially remained unchanged. When fast digital controls were not available, it was critical that the fundamental forces dictated by the laws of physics were harnessed. Our engineering forefathers should be applauded for a job well done.

The Evolution of the Modern Grid

In 1904, the Institution of Electrical Engineers (IEE) and the American Institute of Electrical Engineers (AIEE) organized a meeting in St. Louis (Fig. 2.2) that was attended by several hundred engineers who started to lay out the basic rules of how various devices should operate to allow interconnection, interoperability, dynamic balancing (remember, no energy storage!), and scaling of a geo-dispersed power system that could grow to meet local, regional, and national needs. Acceptance of standard voltage ranges and frequencies to meet residential, business, and industrial needs allowed equipment and end-use appliances to be developed that could operate anywhere in the country. A "meshed" network topology was developed for transmission of power over long distances to ensure reliability, while a simpler radial system was preferred for distribution of power from substations to actual end users. Fast-acting protection in the form of fuses and breakers ensured safety while also creating mechanisms for isolating faulted sections of the grid, allowing the rest of the system to keep operating while repairs could safely be done.

Perhaps the most challenging element was control of the many generators that simultaneously acted to form and maintain the grid. They had to connect to various types of prime movers, from low-speed Francis turbines driven by low-head hydro to high-speed Pelton Wheels for high-head hydro or steam from coal. Machines

Fig. 2.2 Photograph of IEE and AIEE engineers in St Louis in 1904 formulating grid operation rules (for further reading, please see: https://www.uh.edu/class/news/archive/2023/march/uh-historians-team-up-with-engineers-to-understand-the-future-of-the-power-grid/)

with different "pole numbers" allowed matching of different rotational speeds with the 60 hertz electrical frequency. A synchronization technique that eliminated inrush currents allowed these large machines to gently connect to the system (like a merging lane for high-speed traffic on the interstate). A "power-frequency droop" rule was developed, which allowed generators of different power rating to connect to the grid and to automatically share power with other generators in proportion to their rating and to reach equilibrium with the connected load without requiring energy storage. Because the loads turned on and off whenever they wanted, and because transient power flows were not totally controllable, the idea of running one generator on the system to pick up the slack, or in technical parlance the "slack-bus," was developed. This also allowed the system to keep operating even if one generator dropped off due to a contingency (low latency communication was not an option). Black-starting the system after an unexpected outage involved forming the grid voltage, starting with a single generator, and slowly building the network – a cumbersome and time-consuming process. As a result, the grid was designed with redundancy so as to be continuously operating, even as individual elements failed or were removed for service.

The ability to parallel generators on the grid, and to configure a meshed bulk-power transmission network realized many significant benefits. It allowed the use of larger generation plants, which tended to realize significantly lower capital and operating cost per kilowatt-hour (kWh) of energy generated, simply due to economies of scale. Perhaps less obvious is the impact of load diversity and the fact that all connected loads do not peak at the same time. For a site-located captive generation plant serving a single customer, the plant had to be sized to cover the highest load that it would ever serve. For instance, we may have a peak load of 200 Amperes in our house (if all loads were to simultaneously turn-on), but probably rarely see that peak, if ever. If we had a generator for our entire house load, we would need to size it to provide the full 200 Amperes. If we could combine all the houses in our neighborhood together; because the peak loads for each house were not coincident, the peak power for the aggregated customers could be significantly lower than just the aggregated peak load (this ratio is called the demand factor).

At the system level, this load diversity results in much lower generation and feeder sizing to serve the connected load, especially when compared with a local generation system consisting of multiple generators that are sized to serve the aggregated peak demand. Another major benefit comes from the electromechanical coupling of all the generators. Effectively, the many generators rotate as if they were one massive generator, representing a very large rotational inertia, which provides energy to clear faults and to keep the system synchronized and operating through major transients and disturbances. This rotational mechanical energy is inherently present in the system and was recognized as being critical to maintaining the integrity and stability of the interconnected grid. It should be noted that these modes of behavior represented the intrinsic behavior of the system, as there were no controllers that were fast enough to control the transient and fault mode behavior (other than through tripping a breaker or blowing a fuse).

With basic operating principles enshrined in standard practices, industry could move forward with innovation, addressing issues, filling gaps, and growing the market. Key innovations that drove electrification to national scale include transformers, electric machines, dielectrics, ability to parallel a large number of different types of generators, protection, billing systems, dispatchable generation, and reliability through redundancy (at the bulk system level). Supervisory Control and Data Acquisition (SCADA) systems were developed first in the 1920s, leading to automatic generator controls (AGC) that allowed interconnection and control of multiple generators. As the system expanded, the need for automating control and protection of systems with multiple generators grew.

Along with the increasing complexity of the system came the need to understand what was happening in all parts of the system, and to be able to analyze and predict the response of any action that was taken. This led to Energy Management Systems (EMS), which initially used hand calculations and eventually analog computers to model key aspects of the projected system response, allowing improved optimization of the system, which in turn led to lower operating costs. Improved understanding of the "state" of the system, and the ability to do power flow calculations on the operating system allowed grid operators to schedule and perform economic dispatch (ED) of their generation resources, keeping in mind that constraints such as transmission line capacity were not exceeded and that all loads were served. Computations to optimize grid operations were very complex and evolved only slowly, because very rudimentary computation and communication capabilities were available in those days. It is clear that the grid in the early to mid-twentieth century represented an amazing achievement, where in spite of so many technology limitations, through good engineering and strict adherence to processes, the electricity grid evolved into one of the safest and most reliable mechanisms for delivering energy at scale to power all facets of our lives.

The Grid Unleashed

Even as the electricity sector began to grow, now propelled by standardization and continued innovation, many challenges inevitably came into view. The high cost to build and operate electrical generation and power delivery, and the rapid growth in the number of companies wanting to serve an emerging market, created huge challenges in terms of economic viability. Entrepreneurs and businesses had to focus on elite customer groups, such as high-population-density urban communities who could be served at low cost, or wealthy customers who could afford to pay the premium charges that were needed for financial viability. To spread the load over a day, the companies had to expand their footprint by acquiring other service providers and types of customers, helping to create a more diverse customer base, so their plant and equipment could see better utilization. However, economics still dictated that only elite customers who could afford to pay the higher rates could economically be served. As a result, low-density population and low-income areas, in particular rural and farming communities, were generally being left behind in this energy transformation. Yet, the potential benefits of wide-scale energy access were clear and the demand for access to electricity kept growing.

This led to the first government investigation of the electricity industry by the Federal Trade Commission over 1928–35, which led to the passing of the Public Utilities Holding Company Act (PUHCA) in 1935, accompanied by the bitter politics that typically accompanies change that creates winners and losers and has major economic impact. The Federal Power Act was also passed in 1935, giving the Federal Power Commission regulatory power over interstate and wholesale transactions and transmission of electric power. Public utilities were set up as regulated monopolies, typically inside state boundaries, and had the ability to access financing at low rates with the ability to operate within their service territories with no competition, but with an obligation to serve everyone. Utilities were regulated by state Public Utility Commissions who ensured that they fulfilled their "must-serve" energy-equity mandate, but also ensured that they earned a "reasonable" rate of return on their investments and operations. The government also set up the Tennessee Valley Authority (TVA) and Bonneville Power Authority (BPA) in 1933 to build and operate an extensive system of hydroelectric dams to supply low-cost power to rural customers and for industrialization, capability that was also critically important to support World War II efforts. The Rural Electrification Act was passed in 1936 to ensure that electric cooperatives that were formed to supply low-density population rural areas and farms with electricity also had access to low-cost federal loans. These developments finally helped to make electric power widely available across the entire United States (Fig. 2.3) [4].

Spurred by regulations, electricity consumption during this period in the USA grew rapidly from virtually nothing in 1900 to 200 GWHr/year by 1945. The electric grid became the backbone of the economy, both in the USA and globally, and electricity access became a right, with utilities mandated to supply electricity to all customers in their service territory. As a consequence of this growth, we now have over 3300 electric utilities in the USA, including 300 Investor-Owned Utilities (IOUs), 950 municipal utilities (munis), and around 900 electric cooperatives (co-ops). Electricity became a very effective means for delivering energy. However, other than a limited amount of hydropower (limited due to siting issues), virtually all generation was coal-based thermal. Coal plants had their share of problems – high levels of emissions, including smoke, particulates, sulfur oxides (SOX), nitrogen oxides (NOX), and CO_2, leading to issues of smog, acid rain, and high levels of

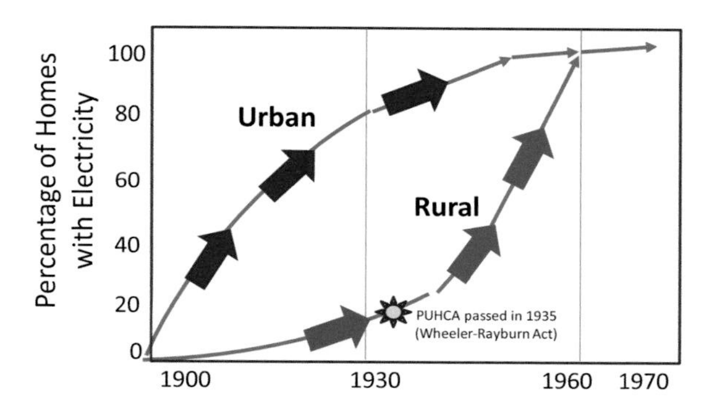

Fig. 2.3 Growth of urban and rural electrification in the USA

pollution. Managing coal ash started becoming an issue because it was rich in poisonous metals which could leach into the environment. Finally, coal plants also had very long ramp-up and ramp-down cycles (4–6 h) and had a slow response to load changes. Mechanisms for energy storage (e.g., pumped hydro) or fast response generation were needed to partner with coal generation.

Regulation also completely changed the "DNA" of the electric utilities – from fast-moving innovative and competitive companies to slow-moving risk-averse entities focused on meeting mandates set by their regulators with little incentive to innovate. The ability to earn a guaranteed return on equipment purchases that were authorized led to a tendency to build large capital projects whenever possible, resulting in bloated infrastructure. There was little incentive to look for innovations and new solutions that reduced capital cost, especially for infrastructure, because it could actually reduce earnings. The monopoly also meant that utilities could provide assurances to vendors that they would purchase equipment for extended periods once the equipment had been validated and approved. This shifted the burden of innovating new products to the vendors, and utilities lost their ability to innovate and think outside the box. Having utility management be accountable to the PUCs created a need to reduce risk, which resulted in a slow and elaborate process of planning, risk mitigation, and standardization. It also created a culture and an environment where new technology could only be adopted very slowly and deliberately. Regulators were focused on equity, reliability, and low cost of energy to the consumers – and utilities had to deliver. As we will see, this utility operating model has a big impact on how our energy future unfolds.

Science, Technology, and Innovation (1880–1945 CE)

In the period leading up to World War II, industrial research organizations dominated research and received significant federal funding. This ramped up dramatically during the war years. The USA government invested in areas of national interest, such as aviation and defense, but let industry drive industrial and commercial developments. Universities were doing some fundamental research but were mainly focused on turning out trained graduates and played at best a spotty and sporadic role in R&D and its translation to economic value.

World War II preoccupied the world from the mid-1930s to the mid-1940s, but unleashed a wave of innovation, so unprecedented that it deserves its own consideration. This period saw rapid growth of many new technologies, including X-rays, vacuum tube electronics, radar, radio, encryption, submarines, aviation (jet and rocket engines), and nuclear fission – representing some of the major scientific and technology breakthroughs. Many of these innovations needed electricity to operate and would not have been possible without the broad availability of electric power. In energy, lack of access to gasoline during the war stimulated the development of syngas from coal and wood that could be used to fuel automobiles and would provide a key chemical process for organic feedstock and fuels. There were no other primary sources of energy that were available at any scale. However, change was afoot. Unlocking the principles of nuclear fission led to the development of the atomic bomb, which helped to end the war but would also help to launch the nuclear power industry.

As World War II wound down, the dominant source of energy supplying the world continued to be fossil based – coal for electricity generation, and petroleum for mobility, transportation, and some localized generation. The only "renewable" resource was hydropower, which represented a small fraction of the energy consumed globally. The only path forward seemed to be to locate more abundant and cost-effective coal and petroleum resources. After World War II, Europe and Japan were rebuilding from total devastation and there was a realignment of global power, with the USA exercising soft power through the Marshall Plan and the formation of the United Nations and the World Bank, to begin to build a world order that could collaborate, trade, and achieve mutual prosperity. With colonization waning and a retreat of Japan from China, new countries including the USSR, China, and India were also emerging as potentially important players on the world stage. Science, technology, education, and industrial development, along with access to abundant and affordable energy were to become key elements in helping to create this new global economy and world order.

Post-War Economic Growth (1945–1975 CE)

As the countries impacted by the war began to rebuild and emerging economies began to shape their own growth plans, it signaled a new era of relative calm and prosperity around the world – a period that the UN has referred to as the Golden Age of Capitalism. Availability of low-cost capital for reconstruction triggered economic booms across the world, lifting the economies of the USA, England, France, West Germany, Japan, and a few other industrializing nations. As standards of living improved, there was a virtual explosion of demand for products and services, leading to rapid scaling of manufacturing, transportation, and trade. This convergence of market demand, consumption, and supply greatly increased the consumption of electrical energy. At the same time, goods transportation, air travel, and personal mobility were rapidly growing, dramatically increasing the demand for petroleum resources. Oil became the critical raw material that would power the growth and expansion of the global economy.

Oil consumption in the USA increased from 6.46 MBls (Million Barrels/day) in 1950 to 14.7 MBls/day by 1970, a 227% increase in 20 years. By 1970, domestic US oil production had peaked, resulting in an increase in the dependence on oil from the Middle East. Oil consumption also increased in other developed countries, with Germany, Japan, France, and others all leaning heavily on oil imports to run their economies. As known US oil fields were depleted, the importance of oil from the Middle East, a politically turbulent area, was becoming increasingly clear. For instance, the Suez Crisis of 1956–1957 was the first of many events that showed the vulnerability of Europe (which relied on the Middle East for two-thirds of its oil at that time) to oil price gyrations. While oil fields in the Middle East had been developed with help from the Western oil majors, the local governments increasingly took control of their domestic oil production capabilities. Oil producing countries including Iran, Iraq, Kuwait, Saudi Arabia, and Venezuela banded together to form the Organization of Petroleum Exporting Countries (OPEC) in 1960 to ensure that the oil majors did not lower oil prices below posted values, and that they maintained control over pricing and production.

In 1973, in response to US support for Israel in the Yom Kippur war, OPEC enforced an oil embargo against the USA and several Western countries, causing an upheaval in global oil markets, with an almost overnight jump in oil prices from $3/bbl to $12/bbl (bbl is a barrel of crude oil containing 42 US Gallons). For the first time, this showed the economic power that governments controlling critical raw material resources could wield. It also showed the advanced industrialized economic powers that they needed to reduce their dependence on and vulnerability to oil. The International Energy Agency was set up in response, and most countries moved to establish strategic petroleum reserves to damp the price fluctuations and to ride through major transients. It is clear that rapid and continuous economic growth had created an almost insatiable thirst for oil that needed to be quenched, and that those who had the oil also had the economic and political power. Oil, perhaps more than any other resource or issue, has shaped the politics and economics of the world that evolved in the aftermath of World War II. The fact that oil is a fossil fuel and is available only in specific locations gives its owners tremendous power over the global economy – not through their scientific achievements and accomplishments, but rather through a coincidence of where the deposits are located. It is difficult to comprehend how a world dependent on energy sources confined to limited geographical areas could evolve into a just, equitable, and prosperous world – and it is perhaps not surprising that it has exacerbated world tensions and the disparity between the haves and have-nots.

From an energy technology perspective, as applied to oil and mobility, this period saw the emergence of deep-water offshore oil platforms, and significant improvements in analysis of petroleum bearing reservoirs, both on-land and offshore. It also saw an increase in the use of natural gas for heating and other purposes, typically piped in from refineries and oil wells on a regional basis. While other sources of oil were known, such as the Athabasca tar sands in Canada and huge oil deposits in Alaska, these oil sources could not compete on a pure cost basis with high-grade oil that, for instance, Saudi Arabia and Venezuela could simply pump out of the ground. The oil embargo of 1973 provided a strong justification to broaden non-OPEC oil access, and the price spikes suddenly made many of these projects economically viable. It also drove countries to reduce their dependence on imported oil, both through accessing other resources and by reducing consumption through improved energy efficiency.

It is difficult to comprehend how a world dependent on energy sources confined to limited geographical areas could evolve into a just, equitable, and prosperous world – and it is perhaps not surprising that it has exacerbated world tensions and the disparity between the haves and have-nots.

For electricity generation, during the 1945–1975 period, coal was still king, and coal reserves were abundant and available in diverse locations. Total US consumption of electricity grew from around 200 TWH in 1945 to 1747 TWH by 1975 – an increase of 870% in energy usage and significantly more than the increase in oil use over a similar period. Similarly, the number of connected households in the USA increased from 40% in 1945 to almost 99% by 1975 – almost ubiquitous connectivity (Fig. 2.3). Thermal generation and hydropower provided the bulk of the primary resource, with some use of heavy diesel for generation on islands and in countries with poor coal resources. The most important new development in electricity was the emergence of nuclear energy.

Building on the work of Einstein, Bohr, Curie, Rutherford, Fermi, and many others, nuclear technology rapidly progressed from the atomic bomb to nuclear reactors for power generation. The first demonstration of "criticality" in a nuclear pile and the ability to control the reaction was in the Manhattan Project in 1942. The first demonstration of electricity generation in 1951 led to the first nuclear powered submarine, the USS Nautilus in 1954. Russia demonstrated a 5 MW fission plant in 1954, with the first US commercial plant generating 60 MW in 1957. The allure of ostensibly free nuclear power that did not generate smog and pollution was strong, and global generation capacity rapidly increased to 1 GW by 1960, and to 100 GW by the mid-1970s. The 1973 oil embargo further motivated countries such as France, Japan, and even the USA, to move strongly toward a fleet of nuclear plants.

From an electric grid operator perspective, nuclear was another thermal generation plant, with a very long ramp-up time and must-run characteristics, which required a high level of flexibility from other grid-connected generation and load resources. The nuclear plants tended to be large, with ratings in the 2000 MW to 3000 MW range, and given their poor efficiency, required massive cooling systems – typically near a large river or the ocean. It is also worth noting that failure of the cooling system in the reactor designs that were being built could cause serious consequences, including release of radiation into the environment. On the other hand, with strong government support, and a strategic imperative to get nuclear power developed quickly, the transition from first proof of concept to commercial units was very short – around 6 years! The role that top-down policy making could play in accelerating the commercialization of a potentially disruptive energy innovation was evident in this case. It is interesting to note that given the vast resources and the deep-domain scientific knowledge needed, this happened without the proportionate participation of private entrepreneurs and tinkerers, as had been the case in the earlier industrialization cycles. Rather, it was government funding directed at major players from the military-industrial complex who were looking for continued growth in a post-war world, companies, and national labs where the deep-domain experts were already working, which created the perfect storm that allowed rapid commercialization of nuclear technology.

Yet, even as new nuclear plants were being approved and built, opposition to them was also growing. Concerns of radiation leakage and safety following accidents and component failures, concerns of nuclear proliferation and access to nuclear technology by rogue nations, and the issue of long-term disposal of radioactive waste were becoming lightning rods for this pushback against nuclear energy. Further, the lingering and residual adverse effects on the health of people impacted by the Hiroshima and Nagasaki bombs were also becoming visible, raising further cautionary flags about nuclear proliferation. Litigation and tighter regulatory processes and requirements led to significantly longer project cycles as well as cost escalations for most nuclear plants, which in turn led to the cancellation of many projects in the 1970s and 1980s. However, nuclear power provided an opportunity for a clean high-capacity-factor energy source that did not depend on fossil fuels, and its lure persists to this day.

The period from 1945 to 1975 also saw continued technology advances on the grid – wider use of EMS and SCADA systems, improved communications and visibility between grid operators, generation, and key substations, increasing use of computers for real-time state estimation, controls and optimizing generator scheduling – all incremental, deliberate, low-risk advances that were slowly adopted by the industry as they were proven and matured. A centralized generation and control paradigm became the norm, and operational processes were refined and became an integral part of the way utilities worked – and they were largely very successful. By now, electricity was regarded as a very stable, reliable, and economical resource, and many considered access to it a fundamental right for all people. Utilities continued to be incented to make large capital investments, on which they could earn a good rate of return, when they were approved by the regulators. There was very little risk to utility operations, provided they followed approved processes and delivered reliable power to their customers at a rate that was approved by the regulators – which in turn assured them a reasonable profit. Investor-owned utilities (IOUs) controlling millions of customers emerged and now control over 72% of electrical energy supplied in the USA [2]. IOUs provided a stable recurring revenue model with a captive customer base and very low risk and were regarded as safe havens by investors.

Science and Innovation (1945–1975 CE)

The period from 1945 to 1975 saw global economic growth and ushered in an era of a rising standard of living and prosperity in the USA and set the stage for unprecedented change. This was despite two wars (Korea and Vietnam), which were fought by the USA from 1950 to 1975. World War II had shown the importance of research and innovation and the impact that it could have on a nation's prosperity and security.

Vannevar Bush, in his 1945 report, "Science – The Endless Frontier," laid the foundation for universities and national labs playing an increasingly important role in a nation's quest for new technology and capability. The Office of Naval Research (ONR) and AEC (predecessor of NASA) were formed in 1946, while the National Science Foundation was formed in 1950, setting up a mechanism for universities to be funded to engage more deeply in research. This was clearly successful: during this period faculty from US universities received 20 Nobel Prizes. Research funding for universities also increased rapidly during this period.

A new rhythm was established in the USA, where blue-sky scientific research was done at universities and some national labs. Programs were created at the federal level to take issues of national interest and to organize directed research to solve key challenges. Industry research labs took advanced scientific concepts developed by universities as well as industrial and national labs and worked to translate the ideas into commercially relevant and viable implementations. Research at US universities was funded by the National Science Foundation and by industry. There was little incentive for faculty to do research that resulted in commercial products. The emphasis generally was on ivory-tower knowledge generation that was not tainted by commercial expectations. There were a few exceptions, such as the Wisconsin Alumni Research Foundation at the University of Wisconsin-Madison, which was formed in 1925 as one of the first university technology transfer offices to help commercialize UW research – but these did not change the general nature of academic research done in the USA.

This period from 1945 to 1975 set the structure and approach used in the USA for basic research and its possible and eventual translation to market. Blue-sky research at universities with federal and industry support was presented in academic conferences that were also attended by leading researchers from industry. In some cases, where the research was also funded by industry, the evaluation process was faster and deeper. Once the results had been validated, and there was interest and a clear path to how such a technology could impact their business, the industry would evaluate and then possibly license the technology (if the terms were acceptable) or develop its own version. At such time, industry would take over the development of products or solutions that included relevant variations of the technology and which addressed all the commercial issues that were involved.

There was a clear bifurcation of activity – where academia typically addressed pure science or knowledge issues with little concern for practical constraints, and industry developed solutions that were commercially practical and viable. Even in those cases where academia worked on practical issues, they often had at best a very fuzzy idea of what the market needed. As a result, the outcomes were not necessarily aligned with industry needs, requiring significant time, effort, and cost to bring about such alignment. This involved a hand-off from academic researchers to key companies (e.g., GE, Westinghouse, IBM, Bell Labs, Boeing, etc.), who then took over further applied research. This became a prototype of how research got commercialized. The process could take decades, and as global competition was still limited, the model worked reasonably well.

At the national and federal government level, this model also seemed to work well as the USA was able to prevail upon its cold war rival – Soviet Union – in the space race and in terms of economic growth. The USA put the first man on the moon in 1969 – a flagship event that established perceptions of American technology supremacy for decades. Those US universities that had access to research funding became a magnet for aspiring Ph.D. students from around the world, and their presence bolstered US research capabilities and the universities that they joined, and provided the trained technical manpower needed by the US workforce.

The Post World War II period from 1945 to 1975 also resulted in developments that would go on to have very high impact and would change the world that we had known for over 200 years. Tremendous new scientific discoveries, backed by funding from the US government funding and leading companies, were achieved in nuclear energy, space travel, semiconductors, and digital computing. The advent of vacuum tubes, a critical component for radio and television, also allowed computing to go beyond mechanical and analog computing to the dawn of the digital computing era, for example, with the UNIVAC computers (built with 18,000 vacuum tubes). While interesting, and actually used for EMS computations for the power grid, this was not easily scalable. These scientific discoveries were all interesting but did not fundamentally change the energy world – just made it a little better. In the case of nuclear energy, we see that it quickly traveled from lab to real reactors, spurred by the promise of clean and cheap energy, and by large levels of funding from the US government. But it should be noted that nuclear energy still depended on a thermal cycle, limited by thermodynamic limits, and nuclear waste still had to be sustainably taken care of, a problem which persists after more than 70 years of operating nuclear plants.

The fundamental game-changing scientific principles that were developing in the 1940s concerned semiconductors, especially those based on Silicon. Operation of these devices was not based on electro-mechanical-thermal principles, but on an understanding of quantum physics. Improved understanding of quantum physics and electronics led to semiconductor devices in the 1940s and 1950s. Scientific research at MIT, Purdue, and other schools, as well as at Bell Labs, with major innovations and contributions by stalwarts such as Ohi, Bardeen, Brattain, and Shockley, to name a few, led to the development of a reliable and reproducible Silicon transistor. At around the same time, the ability of a semiconductor p-n diode to generate electricity from sunlight was noted, eventually leading to the development of solar cells for space applications, with the first solar cells being launched into space in 1958. In addition to being used in analog and digital circuits, Silicon devices were also built to manage electrical power – leading to power transistors and thyristors, with the first high voltage DC link system demonstrated in 1972.

At the time, such innovations were surely perceived as important, but no one could have anticipated the transformative effect that they would have on the world. It took sustained innovation and development on many fronts. The early integrated circuits developed by Kilby and Noyce at Fairchild and Texas Instruments in the late 1950s, and the efforts by Carver Mead and others to formalize the processes for their design and manufacturing, led to the growth of a new industry segment – integrated circuits. As the processes stabilized and manufacturing yields improved, digital logic circuits evolved into densely packed microprocessors and memory circuits. Switching speeds increased even as size shrank, and costs plummeted. Gordon Moore, co-Founder of Intel, suggested that the density of an integrated circuit measured in number of transistors would double every 2 years – a prediction (now enshrined as Moore's Law) that has more or less held for 50 years, leading to devices today that have over 20 billion transistors packed into a single integrated circuit.

These developments, more than anything else, helped to transform the world over the coming decades.

Science and Digitization Change the World (1975–2000 CE)

The next 25 years, from 1975 to 2000, saw the emergence of the current sprawling ecosystem for science and technology development, and its subsequent translation to commercial use. It also saw the globalization of basic science, technology, and engineering knowledge, with many universities emulating the US model, with premier institutions in virtually every country now focused on scientific research along with education. National labs in leading countries tapped into these knowledge streams and allowed them to focus their research on specific national priorities. A network of scientific institutions with global membership sponsored international conferences and published journals that allowed exchange of information and spurred collaboration. The internet became the great leveler allowing access to and sharing of information at an unprecedented pace.

The technology threads that had become visible over the 1945–1975 period, now took deep root, and started maturing. Moore's Law continued to hold and to enable continued broadening of the impact that digitization was having on our daily lives. This period is dominated by advances in digital electronics, and digitization – or the conversion of existing analog and manual business processes to digital form. The emergence of silicon transistors, integrated circuits and memory devices, and their evolution based on Moore's Law, had substantial impact on our ability to do fast computations, data analytics, and data storage. Starting from the 1960s onward, mainframe computers gained in computation speed and memory. Integrated logic circuits and magnetic memory cores led to the development of mainframe computers in the late 1960s and early 1970s and to minicomputers in the mid-1970s. The microprocessor, developed in the early to mid-1970s, would eventually completely change the paradigm of digital computing, moving the world toward powerful microcomputers, laptops, servers, and distributed computing. By way of example, Fig. 2.4 shows the performance of the best supercomputers measured in floating point operations per second (Flops) over the last 70+ years. From 1945 to 1975, the

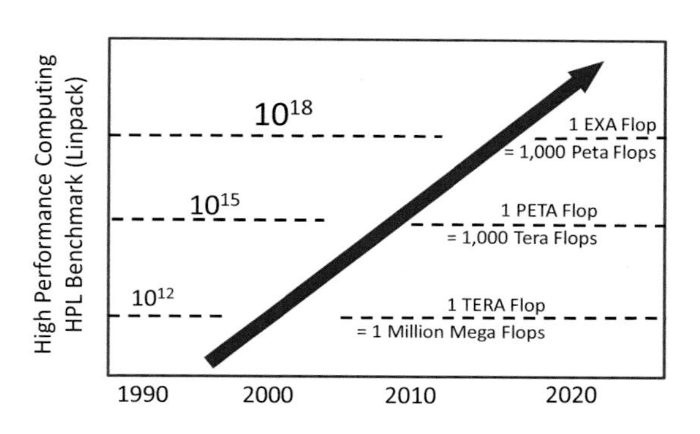

Fig. 2.4 Performance improvements in supercomputers over the last 30 years

performance essentially remained flat (from 10^5 to 10^6 operations), as technology progressed from vacuum tubes to discrete semiconductors and simple ICs. On the other hand, as Moore's Law took hold, we see computing capability increase from 10^6 Flops in 1975 to 10^{12} Flops by 2000 (with continuing growth beyond that) – an unbelievable 1,000,000× performance improvement over 25 years! The same story can be told in ICs, memory circuits, and other related nanotechnologies.

Advanced computing capability had significant impact because it enabled better modeling and simulation of the complex world around us, such as prediction of weather, climate, and ocean currents. It also allowed us to understand the microscopic universe of atoms and particles, to understand material behavior, and to build new materials that did not exist in nature.

This also caused a tectonic shift in the way we innovated. Science now firmly led technology on the path to market, and tinkering-based innovation that was simply driven by personal-experience became more challenging – deep domain knowledge became a key factor for success in technology-related endeavors. On the other hand, as knowledge became global, the fact that someone showed that something could be done, now allowed others to replicate and to improve on those results. This helped to both accelerate and to universalize knowledge and technical capability.

Digitization was a fundamental game changer during this period and beyond, enabling a transformation of our world through digital electronics, simplifying and modularizing complex digital ICs and processors, so they could be programmed at a high level, and it has changed the world forever. It has spawned explosive growth in computing, sensing, communications, telephony – in fact just about everything we use. These technologies, where performance improves on a year over year basis even as cost declines and drives continuous market growth, can be referred to as "Exponential Technologies," and were becoming visible by the end of the twentieth century. These exponential technologies are at the heart of improved electronics, televisions, smart phones, personal computers, medical equipment, the Internet, Cloud Computing, Internet of Things (IoT), video streaming, Zoom calls, GPS, LED lighting, 3D printing, lasers, micromachining, additive manufacturing – the list is endless.

With the advances in computing, machine learning, artificial intelligence, and deep learning, knowledge has exploded, enabling the use of these technologies to improve our understanding of the physical world and to even build things that cannot exist in the natural physical world. We have been able to devise better solutions for health care, agriculture, water, and to build better spacecraft and to make the existing power grid operate in a more optimized and reliable manner. Digital computing and digital technologies have impacted virtually every industry segment, including the energy industry, making it easier and more efficient to do what was done before and allowing us to do things that were simply not otherwise possible.

Every established industry sector has been disrupted or is in the process of being disrupted by digital technologies. Entire new industries have sprung up, industries that could not have existed without our understanding of sciences, technology, and digitization, and eventually digitalization (conversion of existing processes into the digital realm). Yet, the technology risk has been relatively low! Since the late 1990s, the trends have been clear, and technologies have been available to provide the new services and capabilities needed. Moore's Law, with an assurance of lower costs and better performance for core technology elements in the future, guaranteed that the performance, along with profit margins, would improve with time. The cost of implementation and execution has continued to decrease, even as automation has allowed for very rapid scaling with smaller teams – factors that provide immense competitive advantage against incumbents using older more expensive technologies.

This is the trend that Silicon Valley venture capitalists (VCs) have exploited very successfully, creating hundreds of new companies that have completely transformed the world as we know it. It seems they are the masters of innovation and change, and there are no fields where they will not succeed. Silicon Valley has undoubtedly transformed our lives. The VC movement started in the early 1960s and 1970s to provide high-risk capital for new start-ups, mainly in the Bay area. It remained a fairly small group, with $3B in investments in 1983, increasing only to $4B by 1994. The Internet Bubble started in 1995 and ended with the Dot Com boom with $66B invested in 2000, followed by the bursting of the Dot Com bubble just after 2001. The impact of these early investments has been nothing short of transformative and has permeated every aspect of our daily lives.

The Internet and mobile phones (preceded by digital messaging) were a direct consequence of digitalization. The Internet morphed from a service interconnecting government labs and universities to become ubiquitous – growing from 16 M users in 1995 to 361 M by 2000 (continuing to grow to 1.97 B by 2010 and 4.8 B in 2020). Personal connectivity has similarly grown, and virtually every corner of the globe has gotten connected by smart phones, leapfrogging the existing land-based telephone lines that dominated the world for 100 years. Mainframe computers were increasingly replaced, first by minicomputers and then by personal computers – all driven by massive disruptions coming from improvements in digital integrated circuits and digitalization. Lower cost computing, from supercomputers to desktops, and the benefits of improved connectivity – both through the Internet and through mobile phones, began to pervade almost all businesses, making them more efficient and causing wide-scale disruptions to old business models. Companies that did not ride the digitalization wave were losing competitiveness and seeing shrinking markets. As microprocessors developed in capability and dropped in prices in the 1990s, desktops and servers were able to replace functions that mainframe computers provided in the 1970s and 1980s. The Dot Com disaster did not prove debilitating but ushered in a new awareness and capability to continuously upgrade processes through digitalization, which continues to this day.

Changing Energy Industry Structure (1975–2000 CE)

By the mid-1970s, computers and digitization were being deployed in virtually every discipline, including in all aspects of energy. In oil and fossil fuels, improved sensing and computation allowed us to see underground and to tease out where rich oil deposits may be buried, both on land and offshore. Technology was allowing for increased oil recovery from existing oil fields based on 3-D visualization of underground strata. Drilling with intelligent drill-bits and the ability to change direction and do horizontal drilling helped unlock large new reserves that now became economical to access. Improved computer-aided design and manufacturing lowered the cost of building and operating facilities for the drilling and recovery of oil.

Even as these technologies were improving our ability to access oil reserves and to maximize its recovery, the geopolitical landscape was also rapidly changing. The OPEC oil embargo of 1973 continued to have major repercussions, both at the national and global level. All countries responded by trying to reduce their dependence on OPEC oil, promoting increased energy independence and increasing use of coal and nuclear energy. The economic value of oil became clear and triggered a wave of major events including the 1990 invasion of Kuwait by Iraq, followed by the first Gulf War. The year 1991 saw an implosion of the USSR, even as India opened its doors to the world for the first time since its independence in 1947. As Russia privatized its oil industry and received an infusion of technology from the oil majors, it became positioned as a major player in the oil markets. Venezuela also became a major oil player, rapidly increasing oil production by 40% from 1992 to 1998 and refusing to comply with OPEC quotas. Global demand continued to increase at a rapid pace, but so did supply. Concerns about peak-oil never materialized. The late 1990s saw an emergence of the new Asian Tigers – South Korea, Taiwan, Hong Kong, and Singapore followed by China, Indonesia, Malaysia, Thailand, and the Philippines. Even as OPEC opened supply to feed these countries, 1997 saw an economic collapse in Asia, which triggered a slump in the price of oil – a slump that did not recover until the early 2000s. *Overall, petroleum continued to feed the beast of global economic growth, and it did not seem that there was any alternative to oil, that we simply had to live with the political and financial shenanigans that came with a commodity that everyone needed, but which was controlled by only a few.*

At the same time, as the world dealt with successive economic shocks caused by petroleum, the interest in the energy efficiency of automobiles, appliances, industrial processes, and electricity generation increased dramatically – a watt-hour saved was about four watthours (remember the poor efficiency of thermal cycles) that did not have to be extracted from the ground and imported. As a result of digitalization, automobiles changed from purely mechanical machines to smart machines with many embedded microprocessors that improved mileage, added features such as electronic steering, traction control, and anti-lock braking systems, and by the late 1990s, advanced functions such as GPS systems. Digital technologies were coming to the automotive sector. It should be noted that it was in 1996 that GM introduced the EV1, the first modern electric automobile that was designed from the ground up to be an electric vehicle. Electric utilities rallied behind the possibility of electrification of automobiles and began to install charging infrastructure to support the emerging EV industry. However, while the idea was good, the pieces of the puzzle were not all there, and the initiative flamed out. The lead-acid batteries used simply did not provide sufficient range or life to make the vehicle economically viable. As a result, GM called off production by 1999 and recalled all the EV1s, leaving behind disappointed customers and utility partners and a feeling that EVs had been tried once again (remember the 1900s) and were simply not ready for prime time – and perhaps, it was felt, never would be.

The oil embargo of 1973 also rattled the electricity industry. For instance, it helped to accelerate plans for nuclear plants, with 186 new plants approved by 1978, which could potentially have met a large part of the US need for electricity. In fact, of the 99 GW of nuclear generation in the USA currently available, 95 GW came online between 1970 and 1990 [5]. However, following the nuclear accident at Three Mile Island in 1979, 67 new plant builds were cancelled. The cost and time to build nuclear plants increased dramatically due to increased regulatory burden and community pushback. Without new plants to build, the learning rate was lost, and costs continued to escalate. As a result of price escalations and intense public opposition, no new nuclear reactors were commissioned to be built between 1978 and 2013.

Worried about escalating prices of fuel, the Federal Power Commission of 1920 was subsumed into the Federal Energy Regulatory Commission (FERC) in 1977, giving it broader authority to regulate wholesale sale of electricity, oil, and natural gas across state boundaries. The Public Utilities Regulatory Policies Act (PURPA) was enacted in 1978, as a consequence of the 1973 OPEC oil embargo, to encourage energy efficiency and to develop domestic cogeneration and renewable generation in the USA. Natural gas, which up till then had been used primarily for heating and as organic feedstock, started seeing rapid increase in use for electricity generation. Natural gas was still restricted to regional markets, mainly because transportation required pipelines from the wellhead to point of use, and so limited delivery across long distances and across seas. Despite these restrictions, gas-based electricity generation increased from 200 TWh in 1975 (~10% of total US generation) to around 500 TWh by 2000 (~14% of generation), showing strong growth both in energy generated and as a percent of total electricity consumption [2].

PURPA also triggered a lot of change in the very stable utility financing and operating model that we have briefly discussed above. The impact of distributed nonutility-controlled generation took another 10 years to become visible but showed up in almost every aspect of utility operations. For instance, enforcing the purchase of this distributed energy by utilities at their "avoided cost" also allowed for rapid growth of renewables, resulting in 1700 MW of grid connected wind energy in California by 1992. Several states established Renewable Portfolio Standards in the late 1990s, requiring certain percentages of renewable energy in their systems over a specified timeline. PURPA was followed by the Energy Policy Act of 1992, which began to move the electric utilities from a vertically integrated structure to a deregulated system that decoupled generation from power delivery. Independent Power Producers (IPP) were allowed to form and operate competitively and were given permission to connect to transmission lines. Electricity system operators had to put in place market mechanisms and a transparent price exchange. This encouraged IPPs to set up generation plants in locations where gas pipelines and transmission lines intersected (their lowest cost points), independent of whether that was the right point electrically for such an interconnection.

At the same time, as was their job, regulators were also pushing utilities to reduce cost, provide universal access, and improve reliability. The old management adage is "What gets measured gets improved." For the IOUs, it was the everyday metrics of percentage of people connected to the grid; high-reliability and low-downtime of power; and low cost of energy. Remember, unlike normal competitive companies, utilities do not earn profits on the electricity (product) they sell, but rather on the investment in assets (such as generation, substations, lines, etc.) that are used to provide the service. Given the strong growth in electricity demand over this period, regulators approved massive capital investments to help improve their key metrics – improving reliability, keeping costs low and improving electricity access. IOU stocks were perceived to be very safe investments because they had a captive customer base, did not take on risk that would reduce economic returns (there were no competitive pressures to do so), and enjoyed virtually guaranteed returns by providing a dividend in addition to value from stock price escalation. This period from 1975 to 1990 was the golden period for vertically integrated IOUs. Access to electricity covered more than 99% of the population they served. Electricity costs for US customers were among the lowest in the world. System reliability was measured on the bulk power system and was routinely 4–5 nines (99.99% to 99.999% uptime, or 6–60 min downtime in a year). What could go wrong?

But the easy access to capital and the ability to earn a guaranteed profit (easy to achieve when there is strong growth in demand) also meant that utilities did not have to be as efficient with their capital as other capital-intensive industries in competitive markets, such as mining, metals, or oil and gas. Remember also that these investments were all long-horizon investments, with assumed productive asset life spans of 30–50 years over which the economic viability had been assessed – perfectly reasonable for an industry where change occurred slowly and in a managed way. Further, as R&D was not part of what was measured, most utilities virtually eliminated any internal research capability that they maintained, relying instead on vendors to do the research and product development. As industry structure changed, and as technology change accelerated, this investment and operation model would leave utilities vulnerable in the days ahead. The impact of PURPA over this period and the following move toward deregulation would further upset the apple cart.

IOUs, because of their size and capital base, had a large employee base of technical and other staff and had the ability to execute the complex operating rhythm and processes that were part of keeping a large interconnected electrical system, often with millions of customers, operating in real time. IOUs provided about 65% of the electricity delivered to customers – the remaining 35% was provided by municipal public power systems (Munis) and electric cooperatives (Coops). However, from the standpoint of access and geographical spread, it was almost the reverse; 65% of the geographic base of electricity customers in the USA is served by coops and Munis, and not by IOUs. Munis and Coops were started to serve parts of the potential customer base that early electric utilities did not want to serve, because of the challenges of serving them profitably. Munis are owned by their cities and communities and are financed through municipal bonds. On the other hand, Coops are owned by their customers, and tend to be more rural, often with a very small customer base. With government loans through the Rural Electrification Act of 1938, cooperatives were established to serve these rural communities. Today, there are more than 900 Coops in the USA, many serving a very small customer base, and often challenged to maintain staff and technical competency needed to deal with changing technology. The National Rural Electric Cooperatives Association (NRECA), formed in 1942, operates to help its Coop members to manage the trials and tribulations of fast changing policy and technology issues. In 1975, this model was working well, and the electricity industry seemed to be at a good place!

Independent System Operators (ISOs) were eventually formed to manage seven large regions that the US power system was split into, and to provide coordination, power balancing, and market functions. The ISOs did not own any assets but operated under authority vested in them through FERC. As issues with deregulation became visible, and as major utilities and states pushed back, the deregulation process in the USA stalled and was never completed. Today, only 12 states operate with a fully deregulated model, and most others with some form of partial deregulation, adding considerably to the confusion and complexity of managing an interconnected electrical system that spans multiple states.

The above discussion helps to illustrate the increasingly complex, geographically dispersed, and densely interconnected electric grid that was evolving, driven by global currents and politics, federal and state law and policy, and the need to comply with requirements that were often conflicting with each other. Volatile price of generation resources, long delays in building new infrastructure, as well as real-time-physics-based limitations such as generation ramp-rate, congestion on transmission lines, power transfer limits over long lines, inability to store electricity, ensuring stability of large interconnected sub-grids interconnected for reliability, ability to survive faults at the distribution and bulk system level, and the process of restoring power to large regions after cascading large area blackouts – all added further layers of complexity to how utilities operated and the grid was managed.

The 1965 blackout of the Northeastern United States, which left 30 million people without power, showed the fragmented nature of the industry, and the fact that there were no mechanisms to coordinate and address research and development to support their technical and operational needs. This led to the creation of the Electric Power Research Institute (EPRI) in 1972, under the leadership of Chauncey Starr, by federal mandate to act as a cooperative research organization for the electric utility industry. EPRI was funded by a mandated portion of revenues for all US utilities and provided a cooperative research framework to look at challenging issues and to do the pilots and demonstrations needed to de-risk new technologies for utilities to be able to use. This filled a major gap in the capabilities of the industry, but also further weakened any need for individual utilities to maintain internal research capabilities, as they had no competitive issues to worry about. This also led to further weakening of a utility's technical skill base, especially in new technologies that were not yet deployed at scale on their systems and helped to create resistance to the deployment of such new and seemingly "unproven" technologies. The process of piloting, testing, and deploying new technologies continued to be very tedious and time-consuming, as utilities continued to be risk averse and were still measured on the same "reliability" metrics. This discussion is relevant to the broader energy issues because this was also the period when the world was being rapidly changed by digitization and digitalization: How did a slow-moving regulated electricity industry manage this transformation?

Digital technologies also completely changed the way the electricity industry operated, but because of its regulated and noncompetitive structure, the change was implemented in a more controlled manner. In the earlier part of this period, mainframe computers made it possible for electric utilities to better understand and manage their very complex systems, allowing system-state-estimation and optimal-power-flow calculations at regular intervals (e.g., every 15 minutes) to allow for optimal generator dispatch, the main lever for achieving control on the grid. This helped to optimize system operations, improve asset utilization, and improve system stability.

As microprocessors became available, they were integrated into protection relays, improving safety, and avoiding nuisance trips. They also helped to improve asset management and workflows, so people could be deployed better, especially during contingencies. Communications were set up between key generators, substations, and control centers. Elaborate EMS systems were established to maintain visibility and control over key operational parameters at critical points on the bulk power system. The focus for these investments continued to be the bulk power system, while the distribution system remained essentially unmonitored and unchanged.

In the USA, trillions of dollars of capital investments comprising large infrastructure including thousands of generators, seventy-five thousand substations at >69Kv and over 400,000 miles of transmission and sub-transmission network, all had to be coordinated and controlled. This made system modeling and simulation a complex problem, which opened the doors for academic researchers, who had better access to computers and to graduate students, to become more deeply involved in research to help develop the tools with which utilities could optimize their systems. As the power system hardware changed very slowly, the development of which was under the vendor's control anyway, the focus in power-systems research shifted to digital computing to model every aspect of the system, including transient studies, stability analysis, and the impact of contingencies. Because the software programming could be changed easily, the focus in power systems research shifted away from hardware and toward software and smart devices. All these changes were aligned with preserving the status quo of the investments that the utilities had already made, and looked at ways to improve the system performance, utilization, or cost. By the early 1990s, digital technologies were shown to add value to utility operations, and utilities began a concerted effort to increase the adoption of digital technologies to improve the value they could get from investments in their bulk power system infrastructure. Large vendor companies still played a dominant role in defining and bringing new technologies to market in this very highly regulated industry sector.

Digital Technologies and Electricity (1975–2000 CE)

Increased digitalization also continued the journey of the utilities toward improved visibility over their networks, faster cycle times for control actions, and better system and asset optimization, but retained the basic architecture and operating principles that were the backbone of the system. This paid significant dividends because the high-cost components of the bulk power system – e.g., generation, transmission, substations, and large power transformers – the key assets that defined the availability and reliability of the bulk power system, were mostly fully depreciated. Improved and timely visibility to system operating conditions, and load prediction through market mechanisms, allowed utilities to better optimize generation, reducing cost while improving reliability on the bulk power system. As the electricity sector partially deregulated, efficient market operation for the generators became even more important. The role of the system operator and the 150+ control rooms in the reliable operation of the centralized grid became an important driver. It made sense that the availability of more information would lead to better decision making.

This improved operating paradigm allowed very complex operations to be supported that provided hundreds of millions of customers with reliable and low-cost electric power. These developments further drove utilities to adapt and to succeed in this new environment. This resulted in a significant number of innovations, including more sophisticated EMS and SCADA systems, optical-fiber links for low latency communications, Phasor Measurement Units to assess stability margins, optimal power flow computations, economic dispatch of generators, and the idea of locational marginal pricing (LMP). Technology and research were focused on optimizing the use of available controllable assets, such as automatic generator control (AGC) and network switching, to control the overall network itself. The ability to improve utilization of existing assets became even more important as "not in my backyard" or NIMBY opposition grew to the build of new transmission lines and substations within communities.

The key to high bulk-system reliability is the (N-1) criteria, which requires that the system remains operational at full capacity and that all loads can be restored if any single component or subsystem (e.g., a generator or transmission line) fails. Note that this does not mean no short-term outage should occur, only that the load be restored within a specified time. While (N-1) redundancy and improved visibility on the bulk power system dramatically improved its reliability (remember that was a key metric), visibility and knowledge of the system essentially dropped off downstream of the substation where the radial distribution network started. This was because the cost of monitoring millions of points on the distribution system and integrating that data with grid operations was too high. Utilities were still able to get acceptable distribution system reliability by switching circuits when faults occurred and restoring service as quickly as possible when an outage was reported (typically by phone – true even today in many places). Power quality was discussed frequently in the technical community, but was never an operational focus for utilities, except for the rare problem that was typically caused by a large industrial customer. Overall, this was a stable technology and operating paradigm that existed from the

1970s until around 2000 and could have continued for a much longer time if external factors had not forced change.

Because the physical processing of energy was done with the same large power components as before, the utilities could preserve their investment base, and digitalization was mainly used to enhance operations. This generally left the structure of the power system intact, and by itself did not introduce any major disruptions. Utilities continued to operate in their traditional manner as risk-averse highly regulated entities. The new technology introduction process in utilities was still in sync with the old model that only allowed very slow and deliberate adoption. In fact, the process was stacked against rapid change – the life blood of entrepreneurs and their start-ups. Technology introduction could only be done viably by large incumbents over long periods (utilities wanted that anyway because substantial vendor balance sheets were an indicator of their stability). As a result, rapid changes being driven by entrepreneurs and VCs through digitalization in other fields could not be adopted and scaled along a timeline in the utility sector that was attractive for the cash burn, operating cycles, and eventual profitability of the startups that were bringing these new technologies to market (as the personal experience of the authors validates). Hundreds of new entrants, hopeful that their technology would be adopted because it was better and would improve utility operations, died on the vine because of inordinate time delays that inevitably translated into cash-burn and an inability to retain the team and competencies that gave the start-up their edge.

While most of the change brought about by digitalization was in the application of digital computers and microprocessors to improve grid operations, a second thread related to the actual control of electrical power using semiconductor devices was being developed. The first major development in this field was the silicon thyristor developed by GE in the 1950s – an offshoot of the work being done in semiconductor technology. Thyristors were power semiconductor devices that operated like high power switches that could operate at line frequency (60 Hertz) and could directly control large amounts of power. They replaced mercury arc rectifiers, or thyratrons, which had demonstrated the ability to convert available AC power to DC, to transmit it over a long distance, and then to convert it back to AC power. High voltage DC (HVDC) could transfer power over longer distances than AC could because it did not have to overcome the effect of voltage drop along the reactance (inductance) of the transmission lines. The first HVDC system using thyratrons was built in 1954 in Sweden, with the first thyristor-based system in Canada by GE in 1972. The use of HVDC systems grew rapidly, for the first time allowing transmission of large amounts of bulk power over long distances. This allowed unique new capability – through connection of resource rich areas, such as hydro resources in Quebec with New England; interconnection of Brazil (60 Hertz) and Paraguay (50 Hertz) systems at Itaipu transferring over 6300 MW of power; and controlled power flows between large grids that were not perfectly synchronized (Australia). HVDC was the first technology that could actually process high power levels on the grid using semiconductor devices and ushered in the era of dynamic power flow control using power electronics technology. In the late 1990s, the use of Flexible AC Transmission Systems (FACTS) devices was proposed for the AC grid to directly control electrical parameters such as voltage as well as power flows. Power electronics was again at the heart of this new dynamic control capability.

Starting in the early 1980s and accelerating substantially in the 1990s, digital control of power for industrial processes, based on power electronic converters, was also underway. New power semiconductor devices such as IGBTs and MOSFETs were used to build highly efficient power converters, which were controlled by digital microprocessors. These converters could take electrical power in the form it was available locally (e.g., 480 volts AC at 60 Hertz or DC) and convert it into a form that was needed (e.g., DC voltage, or AC at a different voltage or frequency). Power converters were used to provide precise control of electrical power for motor drives, trains, elevators, and a variety of industrial processes, thus improving productivity while reducing energy losses significantly. Given the pace of digital technology growth in nonutility sectors, power electronics also saw tremendous growth over this period. These new technologies enabled compact power supplies for mobile phones, laptops, and digital devices, as well as uninterruptible power supply (UPS) systems for data centers to achieve reliability levels that could not be provided by utilities at the edge of the power grid.

Power electronics was also critical for the emerging area of photovoltaics and wind energy conversion and grid interface – because they all needed to convert the intrinsic DC power generated by these technologies and convert and supply it into the AC grid. However, these new generation resources were still not economical (except with large subsidies) and were at very small penetration levels, and thus did not figure in the planning cycles of major utilities. As a result, by the year 2000, power electronics was seeing unprecedented growth because it used power semiconductors that used similar technologies and fabrication processes that benefited from Moore's Law, with decreasing costs and improving performance, which multiplied the application opportunities, yielding tremendous learning rates for these products. For instance, the cost of a power semiconductor switch decreased from around $10/kW in the late 1980s to $0.50/kW by 2000 – a reduction of 20X and still decreasing! [2]. However, while power electronics was acknowledged as being an important end-use technology, it was not considered to be relevant for managing power on the grid, except for HVDC and to a much more limited extent FACTS – both considered to be very custom low-volume applications, which did not ride the same learning curves.

Energy at the End of the Twentieth Century (1975–2000 CE)

In this one section of the book, spanning a few pages, we have packed the story of 6000 years of human innovation and achievements in the critical life-enabling field of energy.

Our apologies to those who feel the coverage was too short for a topic as important as this, and also to those who feel it was too long, because others have gone over it better and in even more detail. Our intent has been to look deep into the complex mosaic of energy, especially as it has developed over the last 300 years, and to provide an understanding of the intertwined framework that has evolved, a framework that includes dimensions of science, technology, innovation, policy, economics, resource constraints, and random geopolitical events.

Over much of this period, science and technology have moved relatively slowly over periods of decades and more, allowing time for laws and policies to remain reasonably aligned with the changes that technological innovation was bringing. Until the middle of the twentieth century, science generally lagged technology, and innovation was often driven by market need and was the result of experimentation by people with hands-on experience in a field and did not necessarily result from a deep and fundamental understanding of the underlying scientific principles (which were often not known when the solutions were being devised). By 2000, this had substantially changed. Scientific knowledge now provided the underpinnings of discovery and innovation, and the ability to predict and model/simulate phenomenon and behavior that simply could not be done without this knowledge. Further, an exponentially growing knowledge base and computational capability allowed us to innovate faster and to reduce risk and time to market.

In 1999, as we said goodbye to the twentieth century, we were preoccupied with Y2K (Year 2000) issues and whether the transition to a new millennium would trip the date clocks on the computers that now ran the new digital world. Yes, the world was rapidly changing – but we felt that it was changing for the better. Digital devices, mobile phones, the Internet, video communications, and a world of miniaturization promised new capabilities that would have looked like science fiction just a few years ago. Yes, the world of energy would also benefit from these advances, but at a pace and in a manner that was under the control of the utilities, energy industry equipment manufacturers, and its regulators. True, the geopolitical uncertainty in the world of oil cast a dark cloud on our ability to predict how things would unfold. But the worry about running out of oil did not materialize. As oil became more expensive, new resources which could economically be exploited, became available. Nuclear energy, for all its promise, was unable to deliver the world from its fossil fuel woes.

But there were a few elements of change that were also becoming visible. The Environmental Protection Agency (EPA), launched by Nixon in 1970, to protect air and water pollution, was beginning to have influence. Acid rain and air pollution, a direct consequence of tailpipe and smokestack emissions, were starting to become an issue. California introduced automotive efficiency standards and the California Air Resources Board (CARB) introduced limits on automotive emissions. Because California represented a big market, automotive companies started moving to improve mileage of cars, and power companies focused on reducing noxious pollutants from power plants. New technologies such as renewable energy and electric vehicles were visible at the fringes but were not anticipated to be at meaningful scale in any reasonable timeframe – after all, it had taken us 100+ years and trillions of dollars to get to where we were, and the system was working well – how could that change anytime soon!

There was increasing discussion of anthropogenic greenhouse gases (GHG including CO_2, CH_4, SF_6, and others), and the possible impact on global warming and climate change. The Intergovernmental Panel on Climate Change (IPCC) was formed in 1988 and the Kyoto Protocol, to begin to limit GHG emissions, was signed in 1997. Yet, there was little agreement on the extent or impact of global warming, or the steps to take to mitigate the impact. The economic consequences of actually doing something to mitigate the effects seemed so extreme that it was felt nothing could practically be done about it.

At the same time, the international competitive scene was also rapidly changing. Europe and Japan had built new manufacturing capacity and were competing with the USA. China, until 2000, still lagged in education and high-quality manufacturing. However, things were changing rapidly, and nimble competitors with low manufacturing costs were emerging. To maintain profit margins at levels demanded by Wall Street while still remaining competitive, US manufacturers reduced focus on long-term R&D and increasingly embraced low-cost manufacturing from Mexico and China. Because the only thing China (and Mexico) could offer was low-cost labor, this generally also required transfer of technology and workforce training. Access to low-cost goods allowed rapid market expansion and global economic growth. China itself was a large potential market, and Western companies felt they could establish themselves – after all the Chinese had little indigenous technology capability! Of course, there were concerns, but these were all issues that we were confident we could manage. The early twenty-first century was expected to look similar to the twentieth century, with evolutionary and incremental change – nothing disruptive was anticipated. As we will see, we could not have been more wrong!

References

1. https://en.wikipedia.org/wiki/Photosynthetic_efficiency
2. U.S. Energy Information Administration – EIA – Independent Statistics and Analysis.
3. A timeline of history of electricity (electricityforum.com).
4. Two Centuries of Energy in America. In *Four graphs : planet money*. NPR.
5. IAEA Releases 2019 Data on Nuclear Power Plants Operating Experience. IAEA.

Further Reading

Atkinson, R. C., & Blanpied, W. A. (2008). Research universities: Core of the US science and technology system. *Technology in Society, 30*, 30–48.

Bakke, G. (2016). *The grid – The fraying wires between Americans and our energy future*. Bloomsbury Publishing Plc.

Bush, V. (1946). *Science – the endless frontier: A report to the President by Vannevar Bush, Director of the Office of Scientific Research and Development, July 1945*. United States Government Printing Office, Washington.

Carney, M. (2015, September 29). *Breaking the tragedy of the horizon – climate change and financial stability*. Lloyd's of London.

Chesbrough, H. W. (2006). *Open innovation*. HBS Press.

Cohn, J. A. (2017). *The grid: Biography of an American technology*. MIT Press.

Hamilton, J. D. (2010, December 22). Historical oil shocks. In R. E. Parker & R. Whaples (Eds.), *Routledge handbook of major events in economic history* (pp. 239–264). Routledge.

Olah, G. A., Goeppett, A., & Surya Prakash, G. K. (2005). *Beyond oil & gas: The methanol economy*. Wiley.
Sharma, S. (2014). *Energy: India, China, America and rest of the world (Chapter 8)*.
Sivaram, V. (2018). *Taming the sun: Innovations to harness solar energy and power the planet*. MIT Press.
Smil, V. (2017). *Energy and civilization: A history*. MIT Press.
Yergin, D. (2011). *The quest: Energy, security, and the remaking of the modern world*. Penguin.

Energy in the Twenty-First Century: Projections from 2000 Are Totally Wrong!

3

The Forward-Looking View in 2000: "All Is Well"

At the start of the twenty-first century, no one could have predicted that a series of major disruptions would soon confront the energy industry. To experts in all major energy resources, including petroleum, gas, coal, and nuclear, the world in 2000 looked similar to how it did in the late twentieth century: Oil prices had stabilized for much of the 1990s to around $20–25/barrel, an acceptable price that suggested stability and continued economic growth. The abundance of coal spurred increased use at a pace that had remained consistent since World War II. Despite Three Mile Island and Chernobyl, nuclear energy continued to supply a small but important portion of electricity in many countries, including the USA. The vast majority of energy generation still relied on fossil fuels, as it had for 150 years. All signs pointed to a period of stability. It should be no surprise that the International Energy Agency, major governments, and electric utilities all projected several decades of continued stability and steady growth in all aspects of energy resources, unencumbered by major disturbances to the status quo.

At the same time, the electricity sector also appeared stable. Utility companies were focused on using digital technologies to improve service reliability and long-term planning to reduce risk and avoid surprises. Renewable Portfolio Standard (RPS) mandates in many states, which were mainly driven by PURPA, caused utilities to begin to plan for increased deployment of wind and solar energy, but with a slow ramp-up over many decades. After all, early deployments of wind energy in places like Altamont Pass in California had shown high cost and poor reliability. Early demonstration projects by DOE, EPRI, and partner utilities had shown that PV plants and energy storage systems based on lead-acid batteries were clearly not competitive, due to price and battery life, and needed improvements in reliability to meet utility standards. Prior experience suggested that addressing such concerns and getting these technologies to maturity would take 30–50 years. Finally, nuclear

D. Divan, S. Sharma, *ENERGY 2040*,
https://doi.org/10.1007/978-3-031-49417-8_3

fusion was in a similar position – "just beyond the horizon," where it had been for the last 50 years!

Similarly, the automotive industry also seemed to be in a strong and stable situation. Automakers were focused on making their cars better – improving fuel efficiency, SO_x and NO_x emissions, and vehicle safety. Integration of digital electronics with cars improved all aspects of the automobile, including comfort, safety, manufacturing, operation, and maintenance, and introduced new functionality such as GPS navigation, communications, traction control, electronic steering, LED lights, and ABS braking. But the heart of the car, the internal combustion engine (ICE), remained the same. As with the electricity sector, the automotive sector also experimented with reducing its carbon footprint. The first initiative, as discussed earlier, was the EV1 from GM in 1996, which was not successful. This was because technologies related to batteries, propulsion, charging, and management were simply not ready for prime time (and there had not been a steady stream of innovation driven by market pull).

Toyota launched the Prius, a hybrid-electric vehicle (HEV), reaching volume production in 2001 and showing dramatic improvement in mileage through engine optimization and ability to recover regenerative energy during braking. However, hybrids (especially plug-in hybrids) failed to reach massive adoption because the promise to consumers of potential savings on fuel mattered little next to higher upfront costs. As a result, HEVs did not become a major part of any automotive company's product offerings in the early 2000s. Like electric utility companies at the time, the major automotive companies remained more committed to current technology than investing in a different possible future.

Even as climate change began making headlines, change remained slow and inconsistent. The 1990 report by the IPCC (Intergovernmental Panel on Climate Change) warned of catastrophic climate change due to a continuous increase in CO_2 concentrations in the atmosphere, which it correlated with anthropogenic carbon emissions. Everyone agreed that something should be done, but few parties were willing to take on the economic consequences – a standoff of interests that Mark Carney from the Bank of England has called "the Tragedy of Horizons" [1]. For example, there were many proponents of a carbon tax, but no easy way to move forward. Forcing a carbon tax would result in higher cost, have negative impact on growth, and penalize incumbents with financial and political power. Also, countries striving to grow their economies and bring their people to a reasonable standard of living understood that enforcing reduced carbon emissions would impose a double penalty on them: while the developed countries had polluted and emitted CO_2 on their way to prosperity, the "have-nots" were to be unfairly penalized, slowing down growth and putting them at a competitive disadvantage in the global market.

On the other hand, market leaders in developed countries, under pressure from emerging economies such as China and India, were unwilling to lose their profit margins and competitive edge. Allowing China and India to keep polluting with coal plants, while the developed economies had to absorb the higher cost of renewable energy or carbon capture, also seemed to be unfair. In other words, even though everyone agreed in principle that CO_2 emissions should be reduced significantly, the economic penalty made it very challenging to implement. As a result, there was (and continues to be) a lot of talk, but not much action, certainly as we entered the early 2000s.

At the same time, the search continued for alternative energy sources as viable solutions to the problem of carbon emissions. Nuclear plants provided an obvious choice. However, accidents at Three Mile Island and Chernobyl had raised concerns of safety. Long-term disposal of spent fuel also continued to be a problem that vexed the industry and regulators. From 1977 until well into the twenty-first century, no new nuclear plants were built in the USA. This caused hardship for vendors and drove up costs, making it even more challenging for new plants to be commercially viable. Additional disasters, such as at the Fukushima plant, further eroded the role that traditional nuclear power plants played. As for other non-carbon generation resources such as hydropower, it was evident that most of the potential in the USA was already developed, and new plants were seeing increasing opposition due to the potential for disturbed habitats and environmental impact.

Many within the incumbent oil and coal industry touted solutions such as carbon capture and sequestration (CCS), in which CO_2 emissions from coal and gas plants would be extracted, captured, transported, and stored in underground rock formations and caverns. This has proven to be very expensive and challenging, especially to ensure that no CO_2 would leak out for thousands of years. Not surprisingly, this potential solution has also faced criticism because the actual issue of CO_2 management was once again being pushed onto future generations to handle. It is also not surprising that the additional costs of extracting, transporting, and then storing the CO_2 made the "clean coal" plants economically unattractive. As a result, even though commercial scale pilots have been operating for around 20 years, the technology has struggled to find acceptance.

Further, carbon-free generation by itself was hard to sell. PURPA pushed for the adoption of solar and wind energy and paved the path for many pilots that were installed (including Altamont Pass in CA). As of the late 1990s and early 2000s, alternative generation and utilization technologies that were sustainable and emissions-free were very expensive and performed poorly, as compared to existing fossil-fuel-based solutions that had been honed over 100 years in the market. For instance, in 2000, levelized cost of energy (LCOE) estimates for utility-scale PV generation were in the range of $850/MWh (as compared to $50/MWh for coal), while wind energy was at $320/MWh (for commercial and residential scale solar, the LCOE figures were even higher) [2]. As compared with fossil fuel plants, both wind and PV solar had very poor performance in terms of capacity factor, dispatchability, reliability, and they could not be efficiently integrated into grid operations.

At the time, the IEA and others considered that these renewable technologies, starting from a base of almost zero, could not possibly scale in a timely manner. Success required that, without subsidization, prices would need to come down by 95% in a short span of 10–20 years to be economically viable. The entire endeavor seemed preposterous, as though fairy dust and magic were required to make it happen. If past experience was a guide, a slow ramp-up for renewables was the most likely scenario, possibly taking 100 years. Renewable energy was clearly not a major area of concern for traditional energy players in early 2000.

A more important concern for the world of energy in 2000 was the specter of geopolitical and financial upheaval resulting from unanticipated events and emergencies. The aftereffects of the 1991 Iraq-Kuwait and Mideastern conflicts were still putting a spotlight on the need to control fossil fuel supply. Further, increasing demand for oil from a booming global economy, now including China, India, Brazil, and several other OECD countries, began to push the price of oil up to stratospheric levels. The air of irrational exuberance about continued economic growth was punctured and precipitated the global financial crisis and recession of 2008, which caused the price of oil to collapse. Once again, we saw the negative impact of vital energy resources controlled by a few autocratic nations, on the global goal of achieving steady and predictable economic growth.

This backdrop, coupled with the growth of exponential technologies and the emergence of China as a powerhouse for low-cost manufacturing, were to cause change at an unprecedented pace, and at a scale that was not imaginable at the beginning of the twenty-first century. In other words, the general feeling at the start of the twenty-first century that technology driven change would occur slowly did not turn out to be correct. On the other hand, the continued inability to keep a tight control over geopolitics of oil supply – made the economics and politics of oil ever more challenging.

Energy in 2020: Dramatically Different Now

If we take a snapshot of a day in 2020, a short 20 years later, we see a world that is quite different from 2000. We see ourselves in the midst of a global pandemic and a succession of national level lockdowns. For a few years, travel came to a grinding halt and people began to increasingly work from home. The digitalization of internet, proliferation of digital devices, Zoom calls, and video streaming started to dominate our daily lives.

Low-cost and super-efficient digital communication channels enabled rapid and viral spread of information, allowing specific-interest groups to spread their specially targeted messaging without any fact-checks, leading to the concept of "fake news." The mainstream media was abuzz with numerous echo chambers that rapidly amplified messaging, but were difficult to keep track of. The "cloud" was everywhere and now meant something other than condensed vapor in the sky. Smart phone and online assistants have became our lifeline and personal aide "Jeeves," and we could not imagine living without them. Consumer behavior quickly adopted its new digital avatar. Online shopping became the new normal, and we rarely used cash anymore. Cybersecurity risks and concerns are becoming endemic and are impacting everyone. The sharing economy is seeing explosive growth, and Generation-Z cannot even imagine how we lived back in the twentieth century.

The story in energy is also changing in front of our very eyes, and the landscape is continuously shifting. In a short span of 10 years, utility-scale electricity generation based on wind and solar energy became cheaper than coal and natural gas. The vast majority of automotive manufacturers are rapidly converting their manufacturing to electric vehicles, which are now visible everywhere and are ramping up 60% year over year. Electric drones are delivering packages in both war and peace (hopefully nothing nefarious in civilian life, only the things we want!).

In the meantime, US companies, propelled by a desire to realize a sustained fossil fuel supply chain, innovated fracking technologies, and discovered large natural gas reserves to become a major producer of natural gas. With the high oil prices prevailing at the time, the USA appeared to have isolated itself from the vagaries of the global oil market. The frequency and impact of billion-dollar weather-related catastrophes have dramatically increased in the USA and around the globe. China has become a manufacturing juggernaut and has grown to become the second largest economy in the world (the largest if purchasing power parity is considered).

None of these things were anticipated in 2000, just a short 20 years ago. The last 6000 years of history had shown us that major changes and transformations do not occur so fast. So, what happened? Is this just an aberration and will things settle back to normal, or are things now different, and will we have to adjust to a new normal? What does this all have to do with energy anyway? This is a complex set of seemingly unrelated issues that we need to unpack to really understand what is happening. It all starts with the story of renewable energy.

Unexpected Growth of Renewables

Hydropower provides the original form of renewable energy on the grid, accounting for around 20% of electricity generation. However, there is little scope in the USA to develop new large hydroelectric plants. The real growth story is that of wind and PV solar energy. Over the period 2000–2020, wind energy continued on a very strong growth trajectory in the USA, increasing from 4 GW of installed capacity in 2000 to around 100 GW by 2020 providing around 14% or 500 TWh of energy to the US grid [3].

Growth overseas was even stronger, especially in China and Europe, where offshore wind was also seeing broader deployment and stronger growth. During this period, the Levelized Cost of Energy (LCOE) for wind energy dropped by 89%, from ~$300/MWh in 2000 to an average of $33/MWh by 2020, and as low as $20/MWh if federal incentives are factored in [4].

The rise of PV solar has been even more spectacular. At $850/MWh in 2000, there seemed no plausible way that solar could threaten conventional generation. However, the German government enacted the Renewable Sources Act in 2000, which was operational by 2004 and provided a forward-leaning "Feed In Tariff" (FIT) program to encourage PV solar energy deployment that would pay as much as Eurocents 57/kWh, with guaranteed payments per a 20-year schedule [4]. At this price point, roof-top solar became a very attractive investment that was low-risk because it was backed by the government and would pay a monthly dividend for the next 20 years. As a result, adoption soared. The German target was to deploy 52 GW of PV solar, a target that was reached by 2016, well ahead of schedule [4]. The use of a sliding scale for FIT payments ensured innovation and limited the amount paid out at around Euro 11 billion/year.

European investment in solar, starting with Germany and soon furthered by Spain and other countries, provided a defined and assured market for a standard building block, i.e., PV panels and inverter, that could return power to the grid. The potential demand just for Germany, at 200 watts/panel, was 260 million panels over the next 10–15 years. This provided a golden opportunity for upcoming Chinese manufacturers, who with help and de-risking from their government and some questionable technology access plays, rapidly began to develop competency in this new technology area. Keeping an eye on the bigger and longer-term market and driven by intense internal competition and incremental innovation, they passed on the savings to the customers, triggering steep and continued price declines and a massive wave of adoption. At the same time, technology advances improved the efficiency of the solar cells from around 13% in 2000 to 20% in 2020, all while costs continued to drop. The resulting "China price," as it was called, was very difficult for competitors to meet, driving many incumbent solar cell manufacturing companies in Europe and the USA out of business.

At the same time, the lower prices motivated many countries, including Spain, Italy, China, India, and the USA, to put in place incentives and must-meet targets, causing a veritable explosion in demand, with global cumulative installations reaching 760 GW by the end of 2020, with annual demand surging to 140 GW/year and continuing to increase. Concurrently, as demand exploded, the price of PV solar cells continued to drop reaching $0.85/watt DC and $1.34/watt AC in 2020, translating to an LCOE of $20/MWh for some US utility-scale installations. These costs are well below that of coal or natural-gas-based generation and have caused turmoil as utilities and regulators struggle with how to integrate this resource into their systems. The 20-year integrated resource plans favored by the utilities and approved by their PUCs have gone out the window.

It is not surprising that nobody in the traditional energy industry could have forecast in 2000 that the above scenario could have unfolded. As a result of all these developments, prices of utility-scale PV solar today are in the range of $20/MWh – a 97% decline in LCOE over 20 years. Global cumulative installations reached 760 GW by the end of 2020, with installation in 2020 alone of 260 GW (increasing by 50% over 2019) and continuing to increase [5]. Over 80% of new generation installations by 2020 were based on renewables, with a total installed global capacity of wind and PV solar greater than 2000 GW [3].

The International Energy Agency (IEA), a prominent voice in the global energy dialog, which guides governmental cooperation and action, forecast in 2004 that renewables such as wind and solar would account for less than 4% of global energy use by 2030 [4]. Every forecast that IEA and other agencies have made on the adoption of wind and solar, or on the prices for these technologies, have been spectacularly and consistently wrong (Figs. 3.1 and 3.2). How can governments and the energy industry make responsible, fiscally conservative, long-term plans and investment decisions in the face of such poor predictability and volatility? It is not

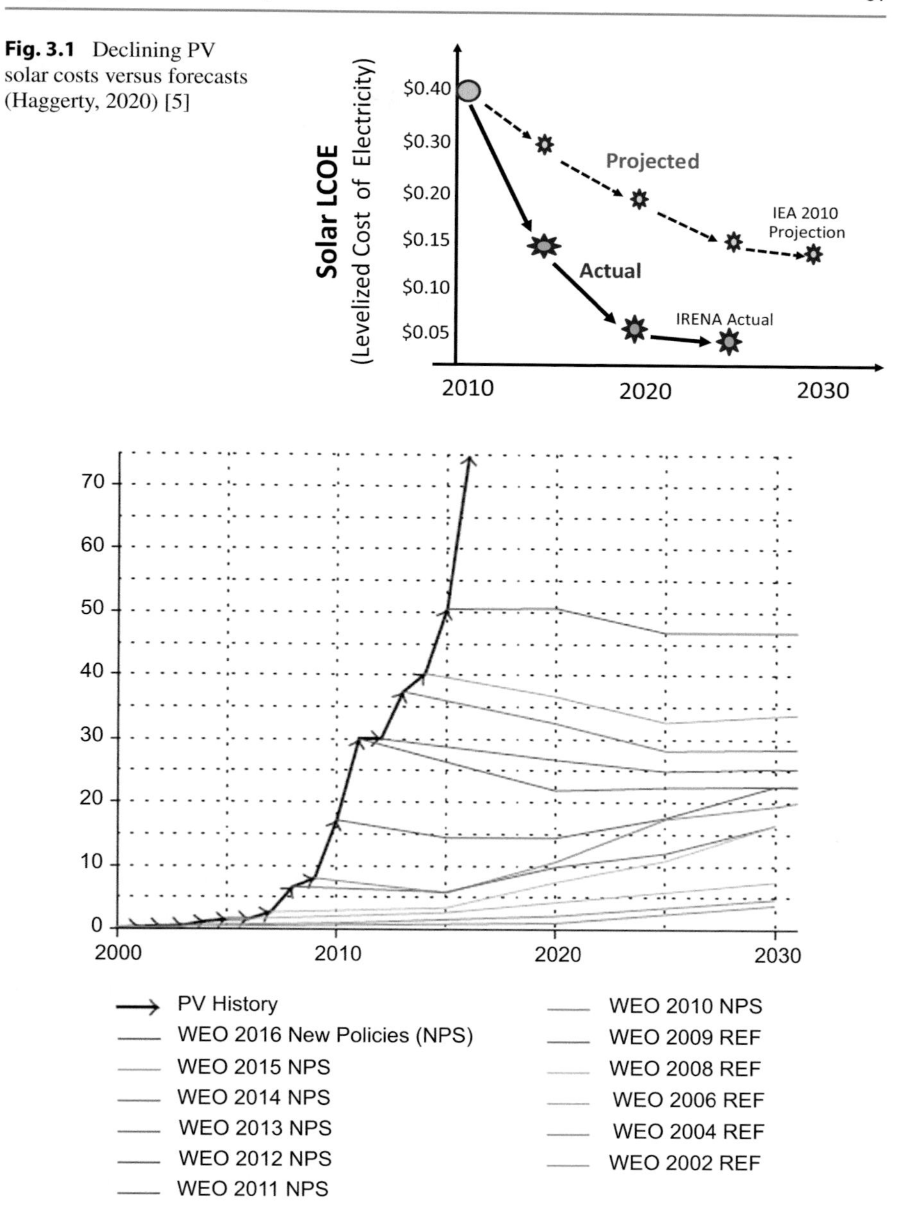

Fig. 3.1 Declining PV solar costs versus forecasts (Haggerty, 2020) [5]

Fig. 3.2 Projections of growth of PV solar in GW of capacity versus actuals (Wikipedia) [5]

surprising that, as recently as 2010, virtually no US utility considered wind and PV as anything other than a nuisance to be managed. Leading US utilities played a significant role in slowing down the adoption of non-dispatchable resources like wind and PV solar, choosing to invest instead in large-scale nuclear plants and carbon capture and sequestration projects. However, nothing could slow down the continued and rapid growth of renewables. This of course had a very big impact on the electricity industry, which was also undergoing its own set of major changes.

Utility Deregulation and Independent Power Producers

Since the late 1990s, many factors had been building up to break up the established monopoly of US electric utilities (e.g., beginning with EPACT 1992 and FERC Order 888 in 1996 and picking up steam in the early twenty-first century with FERC Order 2000). These federal rulings forced many utilities to unbundle generation, transmission, and distribution and to participate in RTOs and ISOs that were formed to manage the overall bulk system, to provide equal transmission access to independent power providers, and to operate energy markets helping drive the costs of energy down (only partial deregulation was achieved and it is questionable how much that helped to drive down costs). However, parts of the USA remained vertically integrated, creating many diverse operating paradigms that could not easily be unified.

In addition, Renewable Portfolio Standard (RPS) mandates, often driven by PURPA, varied widely from state to state, creating high variability in the approach that utilities took in deploying renewables. As the price of energy from wind and solar dropped, utilities had to grapple with integrating a variable non-dispatchable resource into grid operations and energy markets, both of which were designed around the idea of predictable and dispatchable centralized utility-scale generation. By way of contrast, wind and PV resources were generally operated to maximize energy and therefore revenues for the plant owner, which required that plants operate in "maximum peak power tracking" or MPPT mode, independent of what this did to the grid. In the early days, when renewable energy penetration levels were low, the utilities and system operators did not care too much about the variability and unpredictability of this resource. However, as renewable penetration levels increased, major problems surfaced as the dispatchable generator resources had to pick up the difference.

The energy markets (ISOs) operate by computing nodal Locational Marginal Prices (LMP) for a zone depending on balancing of available generation and load, as well as managing congestion and overloading of those transmission lines that can limit desired power flows. Generators place bids on a day ahead basis and are dispatched as needed to meet projected customer demand, while ensuring that all power system assets operate within their physical limits for the current condition and for a set of potential failures. This is a computationally intensive process which leads to the determination of a "security constrained economic dispatch," or SCED, for all generation resources.

Surplus or shortage of non-dispatchable generation, such as excess wind and solar, can cause unanticipated swings in LMPs, resulting in anomalous conditions such as negative LMPs (where you have to pay someone to consume power), that are caused by a juxtaposition of surplus energy in regions and contracts with renewable providers that specify that all their generation must be accepted by the grid [6]. If the renewable generation is curtailed by the grid operator, the return on investment for the plant owner is reduced. The bigger issue is the variability and unpredictability of wind and solar, and the fact that it does not match where and when demand exists. Also, there is no widely deployed capability on the grid today to store energy or to route power from where it is available to where it is needed at a given point in time, including the ability to route around transmission lines that are congested. This results in an inability to dynamically balance the grid and to avoid transmission congestion.

The grid operator works with utilities and market participants at the market, or transactional level, operating in daily, hourly or at best 5-15-minute cycles, with an underlying assumption that the faster dynamics of the system are taken care of by real-time physics-based factors, such as the inertia that is provided by hundreds of rotating synchronous generators and by local protection systems. The automatic and intrinsic response of these generators assures that the grid remains operating and stable, both in the steady state, as well as under all transients and dynamics that can be impressed on the system. This process and way of operating has not changed substantially in over 70 years.

Thermal Generation: Steadily Declining

The net result of the growth of renewables and increasing deregulation also impacted dispatchable generation, most of which was thermal (except for hydropower). Let us start with the facts as they can be observed. As the price of oil surged to $80–140/barrel by early 2008, US shale oil producers were able to use innovative new techniques of horizontal drilling and fracking to access significant quantities of oil and natural gas from many existing wells. This helped to stabilize the price of oil and gas and pushed US generation away from coal and toward natural gas – which gave lower cost and significantly lower CO_2 emissions when compared with the coal plants they replaced. Natural gas generation had typically been used as peaking plants ("peakers") in the past because they could be turned on faster than coal plants could, but now could also be used as a primary resource and replace coal, because of lower emissions, faster ramp-up, faster deployment and now, also a lower cost of base power. The use of combined cycle gas turbines (CCGT) also allowed a significant increase in the efficiency of gas plants, with reported efficiencies of 67%, as compared with 37% for typical coal plants, resulting in lower emissions per MWh generated, as well as dramatically lower waste-heat and a reduced need for cooling.

The economic metrics for gas generation were compelling, and Independent Power Producers could move quickly in a deregulated world, connect to the transmission grid, and begin to play in the market and displace coal. As a result, generation from natural gas increased from 600 TWh in 2000 to 1600 TWh in 2020, while coal-based generation decreased from 2000 TWh to 760 TWh over the same period. As a result, total US emissions from 2000 to 2020 decreased significantly, although global emissions continued to increase [7]. It should be noted that this transition was *not* driven by any climate change concerns, but by pure economics, and alignment with existing policies which allowed for faster adoption.

Nuclear energy (also a form of thermal generation) ran into significant headwinds, in part due to the Fukushima nuclear plant disaster in 2011, which in turn caused countries such as Japan and Germany to redouble their commitment to move away from nuclear energy and toward a renewables-based grid. While nuclear energy in the USA continued to account for around 800 TWh (20%) of US electricity generation through these 20 years, things were not as stable as they appeared. Only two new nuclear reactors were being built during this period, with both experiencing severe delays and cost overruns that call into question their economic viability.

At the same time, 12 US nuclear reactors have shut down since Fukushima with an additional seven shutdowns scheduled through 2025, representing 7% of the US nuclear generation capacity. Most plants are well past their planned life and are operating under extensions that have been granted. Cost, reliability, safety, nuclear waste management, and community pushback continue to be major issues. It is possible that there will be a short renaissance for continuing existing nuclear in some places, but indicators are that traditional nuclear plants will continue to struggle for economic relevance and growth. The net result of these changes is that thermal generation is in decline. This has major consequences as the dispatchable generation that utilities use to ensure a stable grid is no longer available.

Electric Vehicles (EVs) and Energy Storage Accelerating

Another major energy story for this period was the advent of the competitive electric car. As discussed earlier, in the year 2000 the automotive industry did not really consider EVs to be mass market, but rather a niche opportunity for eco-minded customers who could live with compromised performance in return for an environmentally friendly position. The Toyota Prius was released in volume production in 2001 as a hybrid electric vehicle, charging a small Ni-Cd battery when braking and recovering the energy when running. There was no battery available that could meet the requirements for a pure electric vehicle with comparable performance to available ICE cars, and many chemistries had been tried with little success.

Nissan introduced the Leaf in 2010 as perhaps the first mass market EV (they produced over 500,000 vehicles). But as was the norm, it too was what consumers considered a compromise car, with 80 kW of peak power output, a 24-kWh battery, 73-mile range, and very modest acceleration and performance. Unfortunately, although the Leaf has worked well, it too served a niche market and had little impact on the broader automotive industry. The above discussion frames the automotive perspective on electric vehicles circa 2000 – but things were getting ready for massive change!

Tesla Motors, a disruptive Silicon Valley startup, was looking to completely redefine electric cars, positioning them as no-compromise high-performance aspirational vehicles. Their first production vehicle, the Roadster, was launched in 2008 and provided extremely high levels of performance, powering it from 0 to 60 mph in 3.9 s and realizing a driving range of 230 miles. This showed that it could be done, and Tesla followed quickly with the Model S luxury sedan in 2012 and other models, including the Model 3 in 2017. Other manufacturers, especially in China, followed quickly. Partially fueled by a financial incentive for EV purchases in the USA (and other countries), EV sales took off, showing a 60% year-over-year growth, with over 8.5 million EVs on the road today, compared with zero in 2000.

In the period from 2012 to 2020, the automotive industry has gone through gut wrenching change. Every major automotive manufacturer, after pushing back for many years, has now committed to electrification and to stopping the use of the venerated internal combustion engine. This is a sea change that was unimaginable in 2000 (or even in 2010). Certainly, the economic decisions of the automotive companies are driven by commercial realities, including any economic incentives that can help make the transition easier, and not by wishful goals of decarbonization. This should of course not stop EV manufacturers from also claiming bragging rights on the impact they are having on the environment! We would suggest that Tesla certainly deserves a lot of the credit for enabling this transformation. But why did this major transition happen, and how did it happen so suddenly without any advance warning?

The EV story would not have been possible without major developments in batteries. Even though the use of batteries for EVs as a percentage of total battery sales for electronics applications such as mobile phones, started out as a small fraction, it is rapidly increasing to dominate total battery requirements. Projected estimates of the battery market are as high as 1000 GWh of energy storage for EVs by 2025, an unbelievable rate of growth.

The availability of lower-cost batteries that could also handle high power levels saw a reigniting of the interest in grid level energy storage (distinct from uninterruptible power supplies, or UPS, for critical loads such as datacenters, which had been in use for decades). Grid level storage had been tried before by EPRI, DOE, and others using a variety of technologies, including superconducting magnetic energy storage (SMES), compressed air energy storage (CAES), and Vanadium Reflux Flow Batteries, but costs were too high, and these technologies never progressed beyond demonstrations or specific reliability driven applications.

Tesla then proposed the use of the same batteries they used in their EVs to realize grid level storage, providing a first deployment of a 100 MW/130 MWh system in Australia, shaking preconceived notions of the types of energy storage that would work on the grid. Integrated with renewable generation, energy storage was able to offer dispatchability and improved grid integration, significantly improving value, while remaining under the cost of fossil-based generation in many locations.

Growth of energy storage in grid applications started ramping up from virtually nothing in 2000 to about 25 GW in 2020, and is estimated to grow 9X by 2050, to about 215GW. There is still a need for long duration energy storage (LDES) to cover long durations ranging from days to weeks when renewable energy may not be available, which is not covered by these battery energy storage systems (BESS). Pumped hydro is one such resource but, as we have seen, is very limited in terms of availability. LDES continues to be an area of interest and concern, with no widespread solutions available as of 2020.

More Advances at the Grid Edge

These 20 years also saw significant gains in the use of more energy efficient solutions, in many cases by leveraging our newfound ability to efficiently convert electrical energy from the available form to the desired form using power electronics technology. EVs benefit greatly from progress that had been made in the field of motor drives that used inverters (power converters that convert DC to AC) to control the speed, position, or torque of electric motors, initially in industrial applications. EVs also used power converters to control the charging of the EV battery from the grid. As the cost of power converters decreased, they were applied more broadly, including in commercial and residential applications, such as ceiling fans, air conditioners, and refrigerators. Similarly, the need for low voltage DC power for a host of electronics appliances, such as laptops, tablets, and smart phones, saw the development of efficient and highly compact power converters (such as the USB wall port that is universally used), functions and capabilities without which the world of digital electronics would simply not work.

Perhaps the most sweeping change was in the area of lighting. Light emitting diodes (LED), based on semiconductor technology, were initially used as a replacement for indicator lamps in electronics devices, using red and green LEDs that were then available. Development of blue LEDs were critical in the ability to realize "white" LEDs as a replacement for incandescent and fluorescent lights. Much like the trajectory of solar, the number of LEDs has grown exponentially even as costs have reduced, triggering a revolution in lighting – LEDs are now everywhere. Today, 100-watt incandescent lamps have been replaced by LEDs that consume 10 watts and last for 20 years (at least in principle!). LEDs also require a power converter, that is typically integrated into the lamp itself, to convert the available AC power into controlled DC power needed by the LED. Also, like solar, government incentives and procurement programs were critical in driving adoption and reducing price along an accelerated timeline, in which usage of LEDs went from virtually zero in 2005 to almost ubiquitous by 2020. The DOE estimates that LEDs have reduced US energy consumption by 1.3 Quads, saving users $14.7 B in energy bills.

Grid Is Still Generally Fine: But Is Beginning to Show Stress

The above discussion looks at the major changes that were tearing apart the world view from the year 2000 of a stable and predictable energy industry sector. Market penetration of each of these sectors was increasing at an unimaginable and unprecedented rate. It should be noted that all the disruptions outlined above directly intersect with the grid – yet all the changes were driven from within their own silos with almost no regard for how they would impact the grid itself. After all, the grid was viewed as an almost infinite resource, and we could not conceive how anything could impact it in the near term. This perspective was also shared by the utilities and grid operators, who viewed these changes as parasitic and occurring slowly over a time frame that would allow them to put in place standards, compliance mechanisms, and new infrastructure to ensure that the grid continued to operate.

The utility view was that renewable energy was growing at a slow and manageable pace. This view was confirmed by analysis from IEA and others around 2004, which predicted that new renewables would not grow to more than 4% of total energy used by the year 2030. At this slow and measured rate of change, utilities believed they could manage to grow their system in a gradual and controlled manner to ensure that the system kept working properly. Similarly, electrification of transportation would create a massive new load and source of revenue for the grid. Early analysis by utilities and national labs did not raise too many alarms. Even at 125 million EVs, the total energy consumed would only be 125 TWh/year, less than 3% of total US electrical energy consumption. Occurring over 30–40 years at the fastest pace projected, this did not pose an obvious problem. It seemed that their traditional slow rate of growth and dealing with change would be sufficient to manage what was occurring at the grid edge.

So, as all this change was happening around it, what was the electricity sector itself focused on? We saw that in the beginning of 2000 (and up until 2010), utilities were fairly complacent and felt that things were generally under control and believed that "All is Well." Remember that the utilities operate with a long range (typically 20-year) integrated resources planning or IRP cycle. Digitalization of the bulk power system, initiated in the twentieth century, promised improved reliability through better SCADA systems, better system level optimization, and coordination of available resources. However, the 2003 blackout that darkened a large part of the US Northeast began to show that better situational awareness and system resiliency, especially at the grid edge, was needed to prevent such cascading failures.

The idea of a "smart grid" gained traction (what's not to like about something "smart"!). Advances in phasor measurement units (PMUs measure the "angle" difference between voltages in different parts of the transmission system to help assess system stability), smart meters, and other digital technologies became a major focal point for DOE initiatives focused on smart grid and grid modernization. The Smart Grid Investment Grants (SGIG) provided by the Obama administration provided over $4.5 billion to accelerate digitalization through deployment of smart meters or advanced metering infrastructure (AMI) and distribution management systems (DMS) to realize increased visibility into the distribution system, and to improve reliability and service restoration – all primary objectives for utilities. Over 95 million smart meters, representing 50% of connected customers, have since been deployed in the USA. But somehow, most distribution networks remained "dark" and without any level of dynamic control, providing poor visibility and controllability for utilities and grid operators. The much-touted benefits of the smart grid remained elusive [8].

True, growth of new variable generation and the contracts that forced utilities to accept the energy was resulting in anomalous behavior, such as negative LMPs. There was concern that increasing solar generation on the distribution grid would result in voltage volatility and reverse power flows with safety and protection challenges. There was growing concern that PV and wind inverters would interfere with the dynamics of the grid. New standards, such as IEEE 1547 and IEEE 2800, that specified how inverters would behave under abnormal grid conditions, were formulated to take care of that issue. As cost of renewables approached grid parity, utilities could no longer ignore renewables (given their mandate for low cost) and had to seriously consider integrating PV and wind energy with their generation mix. The bidding process now created conditions that forced the use of dispatchable gas generation only for contingencies and not for base load, thus destroying the economics of the gas plant.

Utilities were beginning to deploy Distributed Energy Resource Management Systems (DERMS) to help coordinate the plethora of Distributed Energy Resources (DERs) that were now being deployed and were operating with little to no coordination. Loss of thermal generation on the grid was also resulting in changed behavior of the grid itself, but no one was quite sure what this would entail. But, even in 2020, with the exception of a few grid operators in Europe and Hawaii who had begun to live with the consequences of high renewable penetration, most US utilities remained oblivious to what was coming down the pike. They were not yet seeing the issues on their own systems and were content to focus on today's problems.

There was another elephant in the room. New Orleans suffered major power disruptions that lasted for many months following Hurricane Katrina in 2005. Hurricane Sandy in 2012 brought New York City to its knees, causing 44 deaths and $19B in damage. Puerto Rico saw up to 240 days of power outages for some communities after Hurricane Maria in 2017. California suffered extensive blackouts due to forest fires in 2019/21 – including for many residents who had PV panels on their roofs and EVs in their garages. Texas suffered extensive blackouts in the middle of a brutal winter storm in early 2021. The list is endless and points to a major concern with the way the grid operates today.

Reliability and availability are typically measured on the bulk power system, which is often touted to be 99.9% to 99.999% reliable. At the edge of the grid, where almost all customers are actually connected to the grid, the reliability is much lower. Utilities operating with sufficient visibility and information can work on restoring power following typical small-area events, where a part of the distribution grid is down, but where the bulk system is unaffected. However, with an increasing frequency of "high impact low frequency" or HILF events that also impact wider areas and often the bulk power system, today's centralized paradigm has few answers. This is an issue of grid resiliency and equity and is becoming a major concern.

How Could Expert Predictions Be So Wrong – and Does It Matter

Over the last two decades, we have seen a lot of changes occurring very rapidly in so many areas, all of which intersects with the grid: PV solar and wind energy are now lower cost than conventional coal and natural gas; EVs now offer lower cost to operate than ICE vehicles; all major automotives have made the shift to electrification in a short span of 5–10 years; and energy storage has also decreased in price by 89% over a period of 20 years. All of these game-changing disruptions were completely missed by the very industry sectors that are impacted the most by the changes. Each of these developments dramatically impacts the world of energy, along with the economics, politics, and geopolitics of energy as we have come to think of it.

The geopolitical and financial clout of the oil and gas industry, along with its global dominance over the last 50 years, has been shaken in a short span of less than a decade. The same can be said for a global automotive industry that is being dragged kicking and screaming into a brave new world of electrified transportation. For the electricity industry, a sector accustomed to slow plodding progress with deliberate investment strategies that span decades, and which operates in a risk-averse, highly regulated environment, it is almost inconceivable that such dramatic change would even be allowed. At the same time, international agencies and governments that represent trillions of dollars of global spending on energy and infrastructure, and that are guided in their strategy and investments by the world's leading experts, could not predict any of these trends. At the very least, this has resulted in significant confusion as all these industry vertical segments of Power Generation, Transmission & Distribution, Storage, and Consumption (i.e., Oil & Gas, Utility, Automotive, and Lighting) face major disruption. Perhaps more important, this has led to trillions of dollars in misdirected investments, a result of strategies that did not, or could not, look beyond the incremental change that had been the norm for the last 50+ years.

The strategies pursued by these massive economic industry verticals clearly are a result of thinking and planning by some of the world's leading organizations with deep capability and experience, along with some of the top experts providing guidance. Despite this expertise, these companies as well as organizations such as the IEA and US-EIA, have consistently underestimated the growth rates stemming from a convergence of a set of different and seemingly benign new industry verticals of PV, wind, EV and battery storage. Further, they seemed to be unable to plan and execute on the changes that would be needed on their existing systems to accommodate the most recent projections. As is true in the midst of a major disruption, most incumbent companies and the experts (who are typically experts on the systems being disrupted) could not forecast or fully understand the implications of the new future that was rolling out in front of their eyes.

It is well known that the incumbents have a hard time with disruption. But it is amazing that the disruption is accelerating and is occurring rapidly across a broad range of technologies – energy, communications, computing, medicine, finance, manufacturing – and all at the same time! Is it just a coincidence or are there underlying common factors? In the case of energy, has most of the change already occurred and will we now go back to a more normal and measured pace of change, or will this frenetic pace of change continue? Unless we understand the underlying factors that are driving the disruptions, how will we evolve a strategy to manage the change that is occurring? Do we have a view of where we want to go, or are we now simply in a mode of responding to the latest disruption that comes our way? The next chapter takes a more detailed look at the factors driving the disruptions, their interconnectedness as well as interdependencies, and actions that have been taken by the incumbents. Hopefully that will provide us with an improved understanding of why things are evolving the way they are, and to begin to define how and if we can get to where we want to be!

References

1. https://www.youtube.com/watch?v=V5c-eqNxeSQ, Carney (2015).
2. LCOE (Levelized Cost of Energy) in Year 2000: https://en.wikipedia.org/wiki/Levelized_cost_of_electricity
3. USA Today. (2020, April). A 'Wow' moment: US renewable energy hit record 28% in April. What's driving the change?
4. PV Magazine. (2022, July). Germany raises feed-in tariffs for solar up to 750kw.
5. IEA. (2004). Note: More recent reports give a sense of the enormous disparity between these earlier projections and reality. Recent estimates predict that solar installations will triple by 2027; one study finds that the overall market for decarbonization technologies reached $905 billion in 2022. See Stevens, 2022; Pitch Book, 2022. File: Reality versus IEA predictions - annual photovoltaic additions 2002-2016.png - Wikimedia Commons
6. Malik, N. (2022, August 30). *Negative power prices? Blame the US grid for stranding renewable energy*. Bloomberg.
7. https://www.statista.com/statistics/184319/us-electricity-generation-from-natural-gas-since-2000/
8. https://www.energy.gov/oe/recovery-act-smart-grid-investment-grant-sgig-program

Further Reading

Christensen, C. (2012). *The innovator's dilemma*. Harvard Business Review Press.

Cohn, J. A. (2017). *The grid: Biography of an American technology*. MIT Press.

Enkhardt, S. (2022, August 1). Germany deployed 3.8 GW of PV in first half of 2022. *PV Magazine*.

Gryta, T., & Mann, T. (2020). *Lights out: Pride, delusion, and the fall of general electric*. Houghton Mifflin Harcourt.

Haggerty, J. (2020). Sunny places could see average solar prices $0.01 or $0.02 per kilowatt-hour within 15 years. *PV Magazine*.

International Energy Agency. (2004, October). *World energy outlook 2004*. (Flagship Report). IEA.

PitchBook. (2022, September 8). *Carbon & emissions tech launch report*. (Report). PitchBook.

Shahan, Z. (2017). *IEA gets hilariously slammed for obsessively inaccurate renewables energy forecasts*. Clean Technica.

Sivaram, V. (2018). *Taming the sun: Innovations to harness solar energy and power the planet*. MIT Press.

Smil, V. (2017). *Energy and civilization: A history*. MIT Press.

Stevens, P. (2022, September 8). *Solar installations will triple in 2027 thanks to climate bill, report predicts*. CNBC.

Weise, E. (2022, June 24). A 'Wow' moment: US renewable energy hit record 28% in April. What's driving the change? *USA Today*.

Willuhn, M. (2022, July 7). Germany raises feed-in tariffs for solar up to 750kw. *PV Magazine*.

Understanding Ongoing Disruptions in Energy

4

How Could We Go from "All is Well" to Global Scale Disruption in 20 Years

The period from 2000 to 2020 was clearly a period of change across many fronts, change that was clearly massive, rapid, and very disruptive to the existing energy landscape. These changes were largely driven by a set of technology advances and their convergence (and not so much the geopolitics of access to oil reserves as one would like to believe). The changes were not anticipated and continue to happen today at an ever-accelerating pace (e.g., Fig. 3.2). Do we expect the changes to now slow down, or is something else happening here, where the pace of change will continue to accelerate well into the future? Major industry sectors that provide, distribute, or use energy – including generation, delivery, oil and automotive, have all either been heavily disrupted, or are facing that challenge over the next decade.

Why did this happen, and why were we unable to anticipate and plan for it? On one hand: PV solar, wind, batteries, and EVs moved at their own breakneck pace within their vertical silos; but on the other, these were also connected with and significantly impacted electric utilities. At the same time: internally driven change at the utilities was happening at a much slower pace as utilities tried to adapt to these external changes incrementally and reactively. Policies often played a catalytic role in this changing landscape through implementation of forward-leaning incentives and regulations, thus enabling explosive growth of PV, EV, and Storage. On the other hand, policies also frequently inhibited innovation through the enforcement of regulations that were crafted for a different era and time.

Capital infusion, as is generally the case, flowed to where there were near-to-midterm opportunities, either for stabilization, or disruption, of the existing system, and preferably where the technology risk seemed to have been mitigated or was perceived to be manageable. The overall disruption of the energy landscape was driven by the intersection and convergence of fast-moving technologies, policies, lower-cost-manufacturing, entrepreneurial innovation, business opportunities

D. Divan, S. Sharma, *ENERGY 2040*,
https://doi.org/10.1007/978-3-031-49417-8_4

enabled by aging infrastructure, infusion of capital into renewables, and a refreshing new momentum for clean environment – all occurring simultaneously in many places across many parts of the globe.

In hindsight, our world, including our relationship with energy, was already changing by the year 2000, but we just didn't see that yet. Things still seemed to be relatively stable with only incremental and evolutionary changes forecast. However, between 2000 and 2020, many of the vectors that had started to become visible over 1975–2000 got connected and converged with each other – setting up a pace of change that was unprecedented and not well anticipated. The fast pace of change forced utilities, industry, and policy makers to *react* to change after it occurred, as opposed to plan for it and manage it.

Innovation clearly played a major role in these changes because technologies in wide use in 2020 seemed to come out of nowhere, on timelines that would have seemed incredible just a few years ago. Who was doing the innovation, how was it making its way to market, and how was it scaling so quickly? Were these innovations fundamentally groundbreaking, or were they incremental with reduced risk? What worked and what did not? How would these changes impact the existing system? Has something changed fundamentally, or is it just an aberration and are we now back in control? While our energy infrastructure has held up well so far, cracks are beginning to appear and raise concerns for the next 10–20 years. To understand this fully, we need to look briefly at digitalization to see how it was a catalyst for the disruptions we see in the energy field.

Digitalization, Learning Curves, and Energy

To understand the accelerated pace of disruption in the field of energy, it is important to understand what was happening with the ongoing digital revolution and why it has been sustained for over seven decades. Digitization was already underway from the 1970s onward. Looking from a high level, it may be hard to differentiate between what happened from 1975 to 2000, and what happened in the years after that. Part of it was the transition from digitization to digitalization. While digitization refers to the act of making analog information digital, digitalization is all about moving existing processes into the digital domain. From 2000 onwards, the trend has clearly shifted toward digitalization, and is an underlying theme for much of the change that is occurring.

As we lived through the past 20 years, life seemed to progress without any major discontinuity (except perhaps for the Covid pandemic starting in 2020). But, even at an individual level, our lives have changed a lot, mostly as a result of digitalization. Perhaps the most significant changes are the prolific use of the smartphone and the internet, which went from a means of basic communication to an all-pervasive mechanism to talk, connect, confer, search, buy, bank, entertain, photograph, and manage information in every facet of our life. Our social life, our family, our friends, our communities, and our "tribes," all interconnected in some way through the mobile devices and the internet (including video streaming and social media, not to mention a deluge of spam and "fake news"). This was a direct outcome of digitalization.

Outside the personal sphere, digitalization has changed virtually every business sector, including healthcare, energy, education, manufacturing, supply chain, finance, commerce, HR, defense, and government. *Every business sector has seen massive disruption, in some cases through centralization enabled by digital technologies, and sometimes with massive decentralization, also enabled by digital technologies.* Access to information and knowledge, anywhere and at any time and by anyone around the globe, became the great leveler, allowing improvements in process and system efficiency, and enabling risk reduction in significant ways. *Digitalization dramatically reduced the cost to provide services at scale and allowed rapid scaling without having to scale infrastructure at the same rate, thus allowing mega-scale corporations to emerge, while also enabling small suppliers to have global reach.* Software-based services showed a completely new business model, where the incremental cost of providing services was virtually zero. Agile software and over the air software updates broke the cycle of slow deliberate improvements in technology and products. This ushered a new era of rapid and continuous improvements, even for products that we had already purchased!

The various industry sectors related to energy have all of course been dealing with new challenges and opportunities posed by digitalization. At a high level, energy industry operations were impacted by digitalization, in the same way other service industries were – through an improvement of processes and service delivery. On the automotive front, digitalization moved the car beyond a means of transportation, adding hundreds of microprocessors and digitally enabled advanced functions such as GPS navigation, entertainment, communication, and safety features. Electric utilities began to extensively use computers for state estimation and dispatch, as well as for billing. And oil majors used advanced seismic analysis to find new oil reserves. The world uses PCs, smartphones, digital networks, and computers to run their complex businesses, and for their people to become more productive and to communicate with each other. Continuous advances in digital integrated circuits and computation capability enabled by ever more advanced processors and memory are the fundamental driver for digitalization, a process that has been advancing since the 1950s.

As discussed in Chap. 2, advances in semiconductor fabrication technology played a major role in helping digitalization reach where it is today. Moore's Law [1] captured that rapid pace of technology advancement beautifully by showing a doubling of the number of devices, and thus the density of an integrated circuit (IC), every 18–30 months. We will look more at this trend in greater detail later. But at this point, it is timely to recognize that the higher densities required increasingly finer geometries to be processed. Transistor sizes of around 1 mm (one million nanometers) in the 1950s had decreased to 130 nm by 2000, and to even finer geometries of 5 nm by 2020 (200,000× smaller). Amazingly, this is now at the scale of multiple atoms! The number of transistors in a chip had increased to 50 billion devices by 2020, and with the advent of vertical multilayered structures, this trend shows no sign of slowing down. Digitalization and Moore's Law could not have occurred without advances in our understanding of quantum physics as well as in our ability to manipulate materials at the microscopic, even atomic level. This in turn required development of machines like scanning electron microscopes (SEMs) and Atomic Surface Microscopes (ASMs), and complex machines for ion implantation and extreme ultraviolet lithography. This needed advanced semiconductor foundries costing many tens of billions of dollars in capital investment to build.

An important fact worth noting is that the metric for Moore's Law is technology-agnostic, allowing many different technologies to overlay, overlap, and compete to achieve the overall desired progress. Moore's Law pointed to one of the first examples of an "exponential" technology where the learning rate has been maintained for over 70 years, and with luck may continue for many more years. It is interesting to note that many people have predicted the demise of Moore's Law, usually because a specific technology is reaching a theoretical limit. But in each case, an alternate approach is then found that allows the continued progression, affirming Moore's Law yet again. Further, because the increased density of transistors, and thus their performance, increases at a pace faster than the cost, this improvement has always resulted in an improved performance versus price point that helps to drive down the cost of these digital technologies and further accelerates digitalization [1].

What is perhaps less known is that the same technologies that enabled digitalization and microminiaturization also supercharged the field of Energy. Manufacturing processes and technologies that are similar to those used in integrated circuit fabrication, are also used in the Energy Industry – to build new power semiconductor devices and converters that can efficiently control power flows, as well as photovoltaic solar cells that directly convert sunlight to electricity. Similarly, our ability to now design and build finely detailed new materials at the microscopic, even atomic level, materials that are not found in nature, opens new possibilities in batteries, energy storage, lighting, and new types of solar cells. Many of these new technologies also seem to follow steep and sustained learning curves in terms of performance and cost, much like Moore's Law. But this does raise some questions – don't we have learning curves in normal businesses? So, why is this different?

It's (Not) a Bulk Material World (Anymore)

The growth of a new technology always brings with it the "experience effect," where a combination of experience, emergence of competition, streamlining of underlying supply chains, and leveraging economies of scale, help us to drive along a specific technology "S" curve ("S" curve is a common industry term that reflects the journey of a new technology from early adoption to mainstream in the market). This reflects a "learning rate" that drives down cost of a specific type of product or technology as volume increases, eventually plateauing until a new discontinuity (such as a new disruptive technology innovation) starts a new curve. This is true for virtually all types of technologies and products and has been a primary driver for cost reduction in everything from bulk chemicals to automobiles.

Experience and learning curves have frequently existed within an organization and plateaued relatively quickly, until a change in technology or end goal started another "S" curve. As an example, in the case of Ford, a learning curve to drive prices lower for the Model-T quickly plateaued and transitioned to a learning curve where the focus was now on improving performance (with increasing cost, prices, and revenues), resulting in the Model-A. However, an even more fundamental questions is – do the classical learning curve models completely explain what is happening today?

We believe, what we are seeing is a complex interplay of multiple factors, where their variability, non-linearity, interdependencies, and intra-constitutive relationships are not well established and understood. Learning curves clearly still apply within an organization and for a specific product or technology. Similarly, other factors apply too: global markets that create pull, global manufacturers who try to capture market share, innovators who respond to opportunity, and technology developments that are rapidly shared globally – factors that allow every new discovery to be rapidly replicated and advanced even further at global scale. This is relatively new and differs significantly with how things were at the end of the twentieth century.

At that time, value realized was often linked with materials, whether it was oil, steel, coal, automobiles, plastics, or gold. For value to be realized, materials had to be mined, processed, manufactured, and shipped to consumers – typically more material implied more value. Centralized manufacturing gave economies of scale and was generally preferred, but required massive levels of shipping, both for raw materials and finished products, and led to supply-chain vulnerabilities (that we have lived through during the Covid pandemic). Access to raw materials, especially those that were unevenly distributed or were rare, gave tremendous geopolitical power to the fortunate few, and set up disparities between countries and people. It seemed to be a zero-sum game, and issues of sustainability and climate only made things more complex.

What is perhaps different now is that for the first time we are able to look, not only at bulk materials to do our work, but at the science of unique and novel materials that provide functionality that could not be imagined even a few years back. The late twentieth century gave us an improved understanding of quantum physics, and an ability to manipulate materials at the nanometer and atomic level, capability that simply could not flow from classical science and the average person's experiential base. In the world of integrated circuits and digital electronics, we moved from combining bulk materials to build junction diodes in the 1950s to integrated circuits built using lateral devices on Silicon wafers, with billions of devices packed on a surface that was only a few thousandths of an inch thick, moving further to vertically layered multi-material devices. Increased density of features was the primary driver for packing more devices on a chip of a given area, thus realizing more functionality and higher speeds. At the same time, improved processes realized smaller and smaller features, reaching as low as 5–50 nanometers recently, while also improving switching speeds, yields, reliability, and testability and reducing cost per chip.

An important point to note is that because the chip was active over only a few microns of thickness, the volume of active Silicon material used to realize the desired functionality was very small (with the rest of the material serving as a mechanical substrate). Brand new materials were designed and manufactured: materials that simply did not exist in nature; materials with unusual properties of strength, electrical and thermal conductivity, and the ability to act as catalysts; materials with precise structure that allowed specific atoms and molecules to pass through, while others were blocked. Micro-actuators could now be built that allowed new functionality such as miniscule accelerators, inertial GPS devices, energy harvesting devices, and fluidic logic. Strange concepts such as quantum computing and action at a distance were being validated. Improved understanding at the genome and molecular level were also key in our ability to develop m-RNA based vaccines against Covid19 in an unbelievably short time, saving millions of lives. And this was happening at an ever-accelerating pace all around the world and represents brand new science that was simply unknown in the twentieth century.

Coal Plants Versus Solar Cells: A Look into the Near Future

So, what does this have to do with energy? As an example, let us consider a 200 MW steam turbine and generator that has been optimized over a hundred years, weighs over 500 tons, and over the course of a 40-year life, burns around 12 million tons of coal (which needs to be mined, processed, and shipped in), produces about 36 million tons of CO_2 and over three million tons of environmentally hazardous coal ash, and requires about 120 million gallons of water every day for cooling (or 1440 billion gallons of water over 40 years!). (As a specific example, the USA generates over 110 million tons of coal ash per year!). By way of contrast, over the next few years we will see solar cells that use just 50 kg (0.05 tons) of active material to cover 1 sq. km of solar cell surface and generate 200 MW of power. No fuel is burned, no additional heat is generated, and no cooling is needed, and there are no CO_2 emissions. On the contrary, the 50 kg of active material required will be deposited on an inexpensive substrate (glass or even paper) to realize 20% efficient solar cells at low and ever decreasing cost as technology improves (cell life is a problem that needs to be solved, but that too seems near).

The core components will also (very soon) have very low cost and could be manufactured on-site using 3-D printing and spray painting of active material on plastic or paper sheets, with little material that needs to be shipped [2]. This projection simply represents a continuation of a 23% learning rate for PV solar that has been achieved with a succession of materials and technologies for over 60 years, driving the cost of solar cells down from $100/watt in 1975 to less than $0.40/watt today, or a Levelized Cost of Energy (LCOE) of $850/MWh in 2000 to under $13/MWh in 2022, and still decreasing [2].

There is no traditional learning rate metric that could anticipate a 10,000× reduction in materials (or 1 billion times if fuel and water over the life of the plant is included) needed to generate this level of energy. Also, unlike a coal plant, the PV solar farm will not generate any additional waste heat and will not contribute to CO_2, and thus global warming.

It is amazing that this reduction has occurred at the overall system level, even when the balance of system costs has not come down at the same pace as the cost of PV solar cells, to the point where balance of system and labor costs in today's systems significantly exceed the cost of the solar cells themselves. With the strong market pull and competition that exists today, this focuses attention on the higher cost items (i.e., the balance of systems), unleashing innovation to find new ways to meet the exploding market need and to make sure that it keeps growing.

The comparison shown above provides a stark contrast between twentieth century and twenty-first century technologies and helps to make the point that use *of bulk materials are no longer the sole indicator of value delivered. Rather, custom designed materials (and devices) that build on these "exponential" technologies should be the preferred pathway for the energy industry going forward.*

At the same time, even as we acknowledge the emergence of a new materials-light paradigm, it should be noted that the fast pace of technology change also creates rapid technology obsolescence, and a challenge to business models – where next year's products are significantly cheaper and better than what you offer this year (we already see that with our TVs and cell phones). For the electricity sector, this upends the long-accepted slow process of scientific discovery followed by long development, validation, planning, and deployment cycles, followed again by equally long periods for investment recovery that are allowed under regulatory process – a timeline that has typically spanned 30–50 years.

It is challenging to build economically meaningful business models for power systems that include a solar plant that was built around 2000 to supply energy for 25 years at $850/MWh without any features such as dispatchability or grid support, and to then have it compete with one built after 2020 to supply energy and grid services at $24/MWh, including 4 h of energy storage and full dispatchability! We are also seeing gas-based generation that was designed and built as a base load plant at $50–80/MWh when used at 60% capacity factor, now only able to participate competitively in the market as a peaker plant or under rare conditions when renewable energy is not available, causing a significant loss of market share to solar and wind generation, completely decimating their business model and investment recovery. If this had happened because of an earth-shaking cataclysm, it would perhaps be understandable. But this was only a logical extension of a learning curve that has been visible and in play since 1975, but that no one believed would actually continue on as long as it has. Maybe it is time to learn that lesson and see the future differently.

The use of quantum physical properties and the ability to manipulate materials at the nanometer level to achieve functions that were accomplished using bulk materials opens the door to an almost infinite progression in performance and the ability to continue to potentially reduce cost along an exponential trajectory. This essentially breaks up the old hegemony of a materials-dominant economy – the assumption that bulk materials were at the core of human development and desires, and that was how we generally perceived value. Today, we live in an era when one smartphone has replaced so many other devices, that we can at least begin to understand that less is often more. So how does this shift specifically help our understanding of what has happened in the last 20 years in energy? It is worth looking at the evolution in a little more detail because it may help us understand what should be done in the future, and possibly more importantly, what should not be done. It should be remembered that science and technology are only one of the factors in this energy transition along with manufacturing, policy, and finance playing very important roles in the way that scaling and impact can occur.

Getting Wind to Grid Parity

The path of renewable energy to grid parity was rocky and happened by fits and starts. In 2000, wind energy had by far lower cost as compared to solar but was still much higher than traditional fossil fuel and other resources. As such, it was presumed that there was no chance it could reach grid parity without subsidies or preferential treatment. The earliest wind turbines connected the blades through a gearbox to an induction generator that was directly tied to the grid. This gave very poor control as it had to operate at fixed speed even as the wind speed varied. But in the 1990s, power converters were very expensive ($1500/kW) and prone to failure, and there were few options. This eventually led to the dual fed induction generator (DFIG) system where a fractionally rated power converter could be used with a special induction machine to allow variable speed operation, allowing more of the wind energy to be recovered, and for cost to be reduced. This topology, however, did not work well under grid transients, and a better configuration was needed.

The cost of power semiconductors was dropping, driven by increased adoption and volume in motor drives, UPS, and power supplies for a variety of digital applications. New devices such as Insulated Gate Bipolar Transistors (IGBTs) and Metal Oxide Semiconductor Field Effect Transistors (MOSFETs) were allowing more reliable and robust power converters that could survive major transients and grid faults. Improvements in digital signal processors (DSP) and field programmable gate arrays (FPGA) realized real-time control with sub-microsecond precision, making the task of high-speed control easier. In effect, power electronics capability was riding on the coattails of a steep learning curve in Silicon processing and computing technologies, as well as an exploding volume in industrial and commercial applications and was able to realize a drop in price of more than 20×. However, as power electronics grew in sophistication, it also decreased in price, to the point where it is now only a small (but important) part (<10%) of the overall cost of a wind turbine system.

In the case of wind energy, developments in power electronics also allowed the migration to a "double conversion" architecture, with a separate converter to convert energy from the rotating generator to an intermediate DC bus, and a separate grid-connected converter to return energy to the grid and to provide grid support and services, which led in turn to maximizing the amount of energy that could be recovered from the wind. This also allowed for the development of direct drive turbines, where the gearbox (a source of problems in the field) could be eliminated. This required the use of large diameter permanent magnet synchronous generators with a high number of magnetic poles, which would allow the generator shaft to rotate at the same low speed as the blades, an architecture that is preferred by some manufacturers for the larger turbines now being built.

For wind generators, the amount of energy varies as the area swept by the blades and the cube power of the wind speed. For maximum energy recovery, the blades have to swivel so as to present the maximum area to the wind. At a given location, a larger wind turbine produces much more energy, growing as the square of the diameter, which is linked to its height. This has led to wind turbines increasing in power capacity from around 200 kW around 1990 to around 3 MW for land-based turbines, and to >12 MW in 2020 for offshore wind, with rotor diameters approaching 200 meters! Offshore wind farms, often with hundreds of wind turbines, tend to be much larger, approaching 1000 MW in total windfarm capacity, in some cases. They also tend to realize greater capacity factors (>40% versus 20% for PV) because the turbines are located in places where the wind blows more consistently [3]. This does create additional requirements for converting the turbine power to DC for transmission to a landside substation using underwater cables, where the DC is reconverted to AC to connect to the grid. This can add significantly to cost and complexity, as well as pose regulatory and grid-integration challenges. It can be seen that as turbine sizes grow, the task of manufacturing, transporting, assembling, and maintaining windfarms becomes very challenging, especially for the ultra large offshore farms. The ability to install, commission, and repair offshore wind turbines in places like the North Sea, with rough seas, storms, and freezing conditions, is also very expensive and challenging.

One can now look at wind energy in the context of steep learning rates and declining costs. The biggest advantage so far for wind has been the square law for swept area and cube law for wind velocity. Increasing the height of a wind turbine by 50% adds around 20% to the cost of the turbine but allows 56% increase in power generated. Similarly, locating a turbine offshore in an area with 25% higher wind speed and a higher capacity factor can generate 250% of the energy produced by an equivalently rated land-based turbine. Given that offshore turbines can also be bigger because of fewer transportation limitations and because they are not visible to people, it is easy to see how wind energy has continued on a path of realizing lower LCOEs for the past 30 years. LCOE for wind has decreased from $300/MWh to as low as $13/MWh in 2021, with capacity factors for new land-based installations of 45% [4]. Cumulative land-based wind installations in the USA now stand at 141 GW, with offshore installations just getting underway, showing overall growth rates of 14% per year. Globally, wind installations are at 899 GW, including 25 GW of offshore wind, with Europe and China leading the way in total installed capacity. Wind turbine manufacturing started in the USA and Europe but is now global in terms of technology and manufacturing, with perhaps the largest systems being made by GE Wind, Vestas, Siemens, Suzlon, and Chinese manufacturers including Goldwind and Ming Yang.

Wind has grown to be a big global business, with individual turbines costing over $2.5 million for a 3 MW land-based turbine and much more for even larger offshore turbines. It should be noted that the cost of wind turbines has stayed around $850/kW from 2000 to 2020, although the size and recovered energy have improved dramatically. Serving this market requires deep pockets to develop and manufacture the new technologies, as well as significant logistics capabilities to execute the large projects. Even as newer wind turbine technologies continue to be developed, including floating turbines and vertical axis turbines, it will be challenging for new competitors to become competitive at scale unless they are exploiting a major technology or business innovation. In addition, aggressive competitors who are flush with cash due to fast expanding markets create a high barrier and risk for the start-ups.

It should also be noted that even a large 500 MW wind farm may only have a few hundred or fewer turbines, the advantages of high-volume manufacturing as a cost driver do not really apply for wind. Yet, it appears that by exploiting the physics of ever larger turbines, the industry should be able to continue driving cost of wind energy down, at least for a few more years. By integrating energy storage, they are also now able to offer more value through increased dispatchability for wind farms for better grid integration, while still keeping costs under grid parity. Turbine manufacturers are also developing new techniques to assemble the large rotors and towers onsite, allowing continued growth in the size and rating of the turbines. However, because of the need to grow turbine size to drive down LCOE, it is likely that continued decrease in price will level off over the next few years. However, it is clear that wind will continue to be one of the lowest cost energy sources for the next 20 years, and with the large farm sizes, is particularly well suited for supplementing the bulk power system.

The Journey for Photovoltaic (PV) Solar

The journey for PV solar is perhaps even more exciting, with even steeper LCOE drops than for wind, even in the face of capacity factors that are typically limited to 20% or so. And unlike wind, it shows the true power of steep and sustained learning rates that are driven by exponential technologies. It also shows the global nature of research today that builds on the findings of others, and continually drives the technology toward improved metrics.

Figure 4.1 shows the development of solar cells from 1975 to 2020, tracking the best efficiency cells that were reported by research groups around the world. It also shows several different technologies, starting from single-crystal Silicon to multi-crystalline Silicon, thin film (Cd-Te, i.e., Cadmium Telluride) and newer technologies such as Perovskites, Organic Cells, and Tandem Cells. It shows a systematic increase in solar cell efficiency from 4–8% in 1975 to 25–47% in 2020. The advances have been driven by simultaneously attacking many different parameters and fabrication techniques to gradually improve the performance of the cell.

Distinctly different from wind is the ability to mass manufacture the cells, with high and sustained learning rates as volume grows. In the meantime, technology

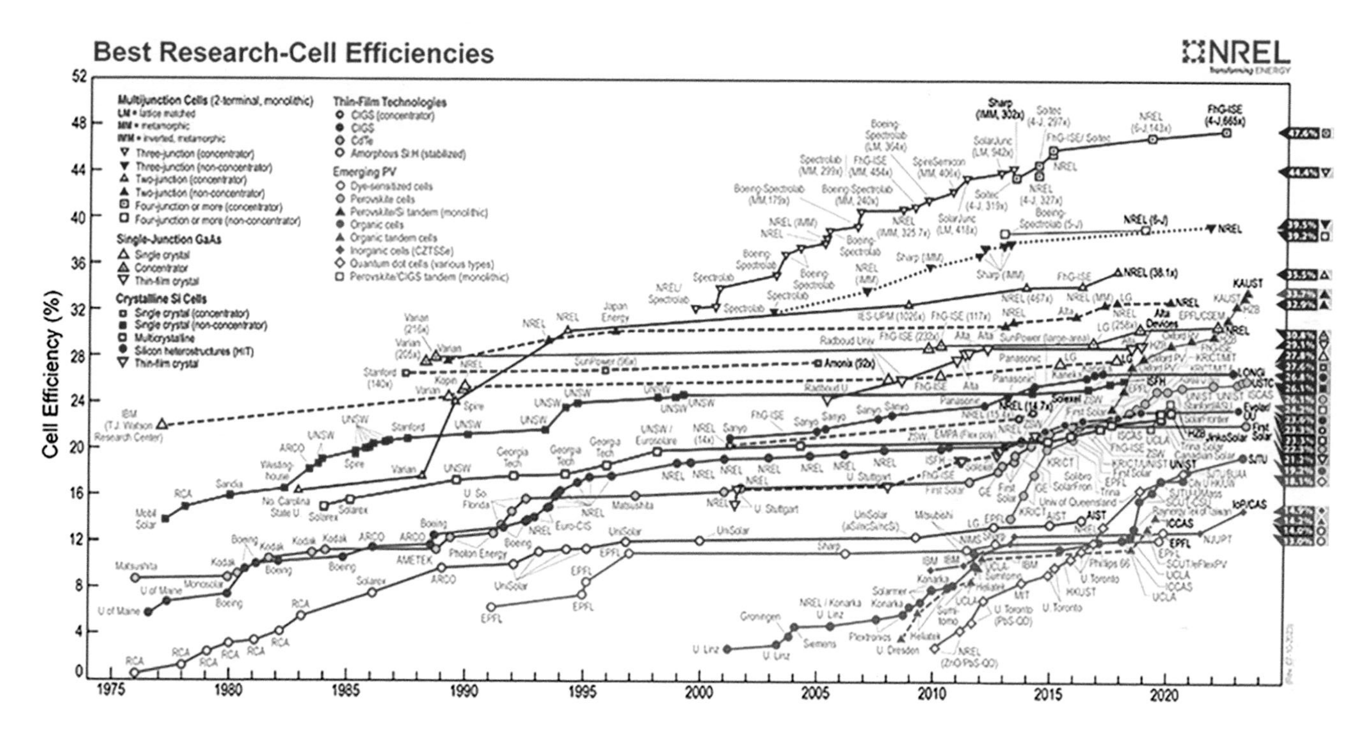

Fig. 4.1 PV Cell efficiencies from 1975 to 2020 (Courtesy: NREL Reports/Website)

advances in related or adjacent technologies start a new learning rate curve. Key factors that have helped to drive the price down, even as performance has improved, are moving from single cell manufacturing to sheet ribbon manufacturing, reducing the temperature at which the cell is made, reducing silver in the panel, and many others. New materials, such as thin film Cadmium Telluride (Cd-Te) and Perovskites provided new pathways along which to continue to drive price reductions and efficiency improvements.

The smallest element in a PV farm is an individual solar cell (typically, measuring ~3 inch × 3 inch though bigger sizes have been tried out). A solar panel may have 50–200 solar cells connected together to create basic modular building block for a PV farm. Individual panel ratings have increased from 100 watts in 2000 to around 500 watts in 2020 (with a panel size of around 3′× 6′). During the same period, the efficiencies of typical production cells have increased from around 12% to 23%. To put things in perspective, a 100 MW wind farm may contain 30 wind turbines of 3.3 MW each. A similarly rated PV farm would contain 200,000 solar panels (made from ~20 million solar cells) – showing how PV solar lends itself to mass production at scale. Also, as the panels are relatively small and can be stacked together, they are also easy to ship globally. It is interesting to note that a 100 MW solar farm occupies around 400 acres of land, versus a wind farm that may occupy between 2000 and 10,000 acres of land or sea surface for offshore wind. However, the on-land wind farm area can easily be used for other activities, such as ranching or farming – something that is more difficult with PV solar.

The biggest challenge is the manufacturing plant. For Silicon technologies, one typically needs a foundry with high temperature clean rooms and long periods of time to process the Silicon wafers. Some thin film technologies possibly have simpler manufacturing processes. Even a modestly sized plant producing 300 MW of PV panels annually needs to assemble 600,000 panels from 60 million solar cells, and to ship them to their customers. In early 2000, most of the PV panel manufacturers were in the USA and Europe, selling PV panels at $850/MWHr into niche applications and pilots. China was opening up and looking to expand its mass manufacturing advantage to the electronics sector. State-of-the-art integrated circuits with fine geometries posed a major challenge as it required technology and foundries that they did not have. They targeted wind and PV solar as key opportunities to scale their manufacturing – to serve their internal needs and take advantage of global opportunities.

With government support, assured internal market access and some questionable methods to get the needed technologies, Chinese manufacturers rapidly developed the capability to supply Silicon solar cells. The internal demand in China and the German feed-in tariff or FIT program triggered intense competition between the Chinese vendors, driving continuous improvements in technology and reductions in price, which in turn triggered increased adoption. Without access to the same public funding support and local market pull, US and European vendors became increasingly challenged to meet "China price," with many of them going out of business. US manufacturers, such as First Solar, who pursued thin-film Cd-Te technology, have managed to stay competitive, but represent a relatively small piece of the market.

It appears that technologies and innovation developed over 30 years around the world, combined with the rapid rise of Chinese mass manufacturing capability, and FIT incentives that allowed commercial breakeven in an increasing number of use cases, created the perfect storm where decreasing prices from the steep learning curves, drove rapid volume growth, which in turn drove further declines in PV prices, which in turn further expanded the market. This combination of events resulted in an astounding decrease in LCOE of PV solar from $850/MWh in 2000 to <$20/MWh today, well below 'grid-parity'. At the solar cell level, this reflected a price decrease from $100/watt in 1975 to less than $0.40/watt today (Fig. 4.2). Further, with ongoing continued improvements in solar cell technology, including completely new technologies and manufacturing processes, and a global surge in demand for PV solar, opens up the possibility that new competitors will emerge, and the business of manufacturing solar cells and panels will become global once again.

While the above discussion has focused on solar cells, there are many other elements to be considered. The panels produce DC power at a relatively low voltage and have to be connected together in series to get to a higher voltage, which is then connected to the AC grid using an inverter, another technology that has been following a steep and sustained learning curve (Fig. 4.3). The rest of the system – including DC wires, structural racking elements, transformers, substations, and power lines to connect to the grid, design and permitting, civil works, etc. – represents the "balance of system" or BOS costs, costs that are generally not following a steep learning curve, and where more innovation is needed. The same solar cell

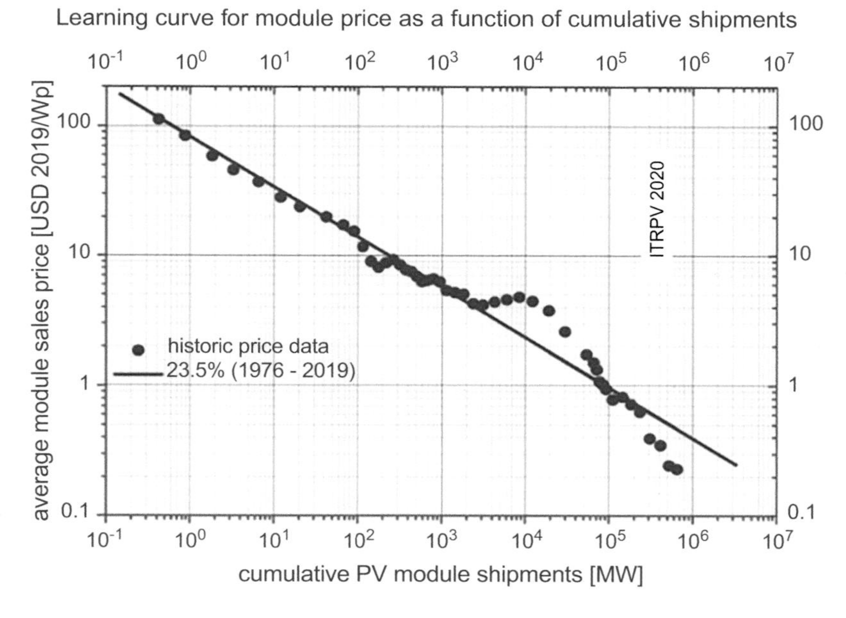

Fig. 4.2 Steep learning curve for PV solar from 1976 to 2019 (https://www.pv-magazine.com/2020/04/28/)

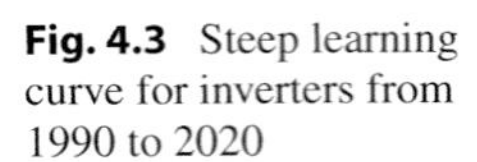

Fig. 4.3 Steep learning curve for inverters from 1990 to 2020

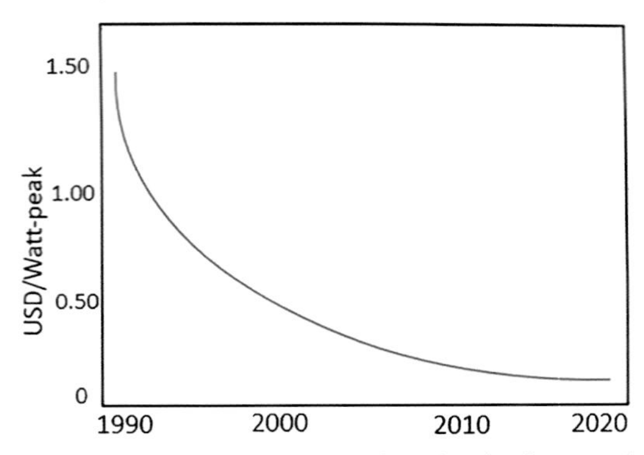

technology can be used for utility scale, commercial scale, and residential rooftop applications – again pointing to the high-volume manufacturing leverage that is now possible.

It is worth observing that until perhaps 2015, there continued to be strong interest in other forms of solar energy. The most important was concentrated solar, where a large array of tracking mirrors focused the sun into a small area, raising the temperature to high values to allow generation of steam. The use of molten salt for energy storage was also developed and demonstrated. It was felt that mirrors with tracking would be lower cost than solar cells (also with tracking at that time), and that the thermal generation used would integrate well with the grid. The steep learning curves for PV solar could not be matched, and concentrated solar has largely been abandoned. This does highlight the fact that with PV solar, as with wind, there is a need to fill in the gaps when the primary natural resource is not there – possibly using an alternate generation mechanism or energy storage.

DOE and EPRI had tried many energy storage demonstration projects in the late 1990s and early 2000s, including batteries based on lead-acid, zinc-air, and redox chemistries, as well as other technologies such as compressed air energy storage – but technological and cost hurdles remained formidable, and a translation to the market could not be made. As a result, both solar and wind energy were deployed without energy storage, letting the grid operator manage grid-integration issues. At low penetration levels, this was not a problem, but as has been discussed before, that issue compounded as penetration levels increased, driven by a surge in deployment that was in turn driven by the steep learning rate. So, why did the energy storage story change all of a sudden?

Where Did the EV Revolution Come From?

We have discussed in Chap. 3 the surprising emergence of electric vehicles, apparently from nowhere. Automotive manufacturers had put to rest the EV revolution in the 1990s. Toyota and Nissan had shown hybrid and pure EVs, but these designs were meant for a niche customer group and would not scale. None of the major automotives were pursuing a fundamental shift to electrified vehicles any time soon. A major constraint limiting the viability of EVs was the nonavailability of an appropriate battery – both from a performance and cost perspective. Many battery chemistries had been tried, including high-temperature sodium-sulfur batteries, but with little success.

The surprising solution to the problem of EVs was found in the rapid ascendance of small batteries for portable electronics and mobile phones. Such devices needed compact light-weight batteries that could last for at least a day before they needed to be recharged and could last for many charge/discharge cycles – nothing reasonable was available. Building on the work by Whittingham, Goodenough, and Yoshino (who would go on to win the Nobel Prize in Chemistry in 2019), Sony developed and began to manufacture the first Lithium-Ion battery in 1991 to power their new Walkman portable music devices, so they would last for a day without having to recharge [5]. By 2005, there were many manufacturers of Lithium chemistry cells, all promising fast charge and discharge, coupled with long life. This was the lifeblood of smartphones, laptops, tablets, and a wide variety of mobile electronic devices. Yet, these were small cells that were mainly intended for low-power electronics rated at the 1–5-watt level. For high-power battery powered applications rated at 1 kW to 1000 kW, such as UPS, forklifts, and battery powered buses, larger form factor batteries were preferred, but did not meet the EV requirements of weight, peak power, energy density, life, and cost.

As discussed in Chap. 3, Tesla Motors completely redefined electric cars, positioning them as no-compromise high-performance aspirational vehicles. Key to this was the rather "insane" idea of interconnecting and integrating 6,831 little "Type 18650 Lithium-Ion Cells" together to make a battery pack. This required very sophisticated packaging and manufacturing, as well as optimization of various aspects such as battery, electrical, mechanical, and thermal management. The battery pack, along with its power electronics inverter and motors, achieved extremely high levels of performance for the Roadster, such as 0–60 mph in 3.9 s and a range of 230 miles. As Tesla displayed the performance of the Roadster, Elon Musk, who was an early investor, took over the company, and launched a broad strategy for ensuring success. This included increasing control over manufacturing as issues related to EVs were different from ICEs, and the need to get the critical technology of batteries under their control.

To achieve that they launched the Giga-Factory to build battery modules to their specific requirements. Tesla used battery cells from partners such as Panasonic and LG that were being used in low-power electronics applications (Lithium Ion 18650 cells), packaging them for structural integrity along with battery and thermal management. The combined requirement led to a dramatic increase in production volume for Lithium-Ion cells and triggered its own steep learning rate. Battery chemistries evolved continuously, leading to learning rates such as we have seen for solar cells. Improved performance parameters for EVs included peak power, energy stored, operating temperature range, increased number of charge-discharge cycles, and reduced flammability. New battery types that are now available include Lithium

Iron Phosphate (LiFePO4) and solid-state batteries, with new developments continuing to occur across the globe at an unprecedented pace. China and South Korea are emerging as leaders in manufacturing the battery cells, although the race is still on.

While the batteries were a key factor, one should not forget the other technologies that made EVs viable. Electric motor drives powered by compact liquid-cooled motors and inverters dramatically simplified the architecture of the vehicle. Peak power levels of 200–500 kW competed with some of the most expensive muscle cars, but now without need for a gasoline engine. Unlike internal combustion engines, which operated with dismal Carnot cycle efficiencies of 20–25% and had to dissipate copious quantities of waste heat to the environment, EVs operated with 90–95% efficiency, and had correspondingly simpler thermal management needs. As the entire engine was now electronically controlled, software integration became easy, and the vehicle essentially turned into a software platform (albeit with power electronics muscle). Further, integration of structural, mechanical, electrical, and thermal design of the batteries and drive train allowed the development of new concepts, such as the "skateboard," which dramatically simplified the design and manufacturing of entire classes of vehicles.

It slowly became clear that, not only were EVs generally higher performance and more fun to drive, but they also cost less to build and maintain than ICE vehicles. Finally, because the cost of electricity was lower per kWh delivered to the wheel than gasoline or diesel, EVs also tend to cost less to operate. It is not surprising that as these factors became clear, most automotive manufacturers understood the coming transformation and the need to move toward electrification. A few automakers, such as Toyota, believed that battery electric vehicles (BEVs) would be limited in range and recharging capability (which we will discuss later), and initially put their focus on hydrogen powered vehicles. It is particularly interesting that Toyota, the company with the longest history with hybrid electric vehicles, felt that battery performance, range, and cost could not progress at the pace needed to make BEVs broadly competitive. They did not believe in the steep and sustained learning curve.

Partially fueled by a financial incentive for EV purchases in the USA (and other countries), global EV sales took off, showing a 60% year-over-year growth, with over 8.5 million EVs on the road today, compared with zero in 2000. It is now anticipated that by 2040, the number of EVs in the USA will increase to between 100 and 200 million, with a mid-level estimate of 125 million by 2035 [6]. In the period from 2012 to 2020, the automotive industry went through gut wrenching change. Every major automotive manufacturer, after pushing back for many years, has now committed to electrification and to stopping the use of the venerated internal combustion engine. Representing nothing less than a sea change, unimaginable in 2000 (or even in 2010), this unprecedented development is driven by the fact that EVs perform better, are more flexible, and cost less to manufacture and operate. And yes, EVs also offer a possible path to decarbonization. But again, while it is advantageous that the new automotive fleet will not emit CO2 when it is operating, the emergence of this technology was at best only tangentially driven by the goal of reducing carbon emissions. Instead, the decisions of the automotive companies were driven by commercial realities, including economic incentives that helped make the transition easier. This should of course not stop EV manufacturers – especially Tesla, from claiming credit for enabling this transformation and positively impacting the environment! But it is evident, as it will be in many other chapters in the story of energy, that economics play an undeniably central role in the transitions that are necessary for sustainability.

Even though the use of batteries for EVs as a percentage of total battery sales for electronics started out as a small fraction, it is rapidly increasing to dominate total global battery requirements. Projected estimates of the battery market are as high as 1000 GWh of energy storage for EVs by 2025, an unbelievable rate of growth. This is in turn driving down the price of batteries from $1500/kWh in 2000, to $300/kWh in 2010, $130/kWh in 2020, with a projected price of less than $100/kWh in 2023–2024, as can be seen in Fig. 4.4. Some EV battery packs are capable of handling discharge and rapid charge at 100–300 kW levels and can withstand 10,000 charging cycles when managed properly. EVs are now projecting about 500,000 to one million miles of operating life, as compared with 150,000 miles for ICE cars.

So, given the earlier discussion about learning rates and supply chains, how did this rapid growth occur? Li-Ion batteries grew from pure science that showed the

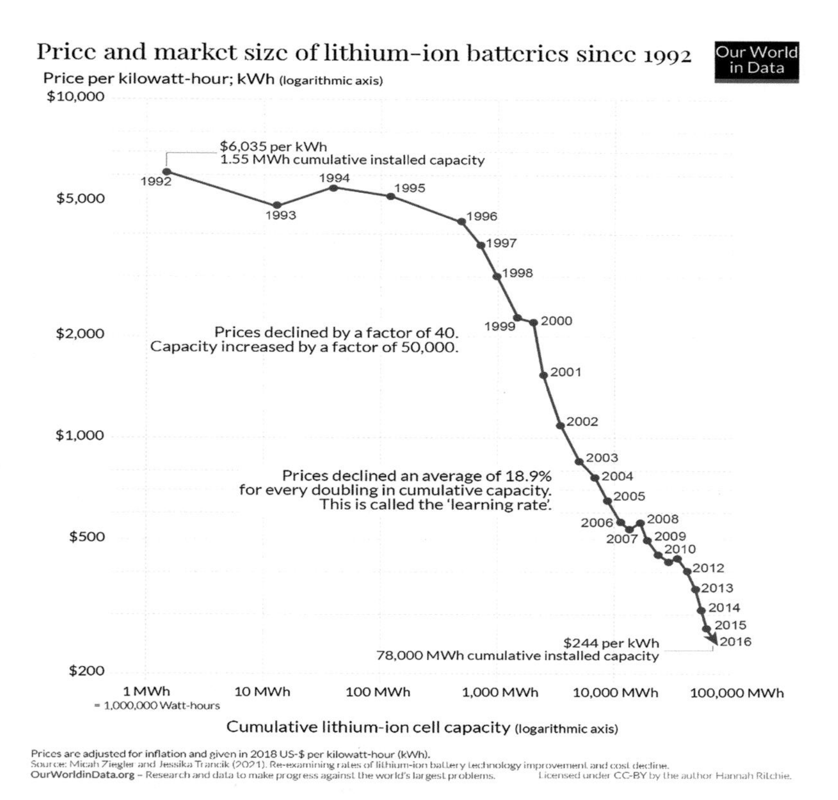

Fig. 4.4 Learning rate and price declines of battery costs (https://ourworldindata.org/battery-price-decline)

possibility of realizing desirable features and a ready market pull from an identified need, with volumes growing quickly because they could adapt existing manufacturing processes that were already supplying large volumes of similar format batteries. The early applications were relatively low power and were viable even with the small battery cell size. There were other battery cell geometries and technologies that may have been better from a pure performance perspective at high power levels, but it would take a lot of investment and time to bring down the learning curve to match the 18650 cells. In the meantime, of course, the 18650 cell itself was riding down a steep learning curve, making that intersection all the more challenging and out in the future.

Tesla knew they would not succeed if they had to create a new optimum battery format, which would have no other customers and instead bet that they could make viable vehicles with the nonoptimal 18650 battery and grab "pole position" in a new emerging industry. It is only now after they have sold millions of vehicles that they are slowly looking to change battery chemistry and format. At the same time, given assured market and volume, many other battery manufacturers have jumped into the fray, initiating the normal frenzied cycle of innovation that marks a steep and sustained learning curve and major transformation. Concerns about whether Lithium is the new oil, and whether Lithium can be recycled are appropriately being raised, and must be addressed, ensuring that we do not create yet another sustainability disaster. On the other hand, the best solution may be to develop battery chemistries that meet the performance requirements and are based on abundant raw materials.

Rise of Grid Energy Storage

As renewable energy penetration on the grid was increasing, it was very clear that there was a need for grid level energy storage. The most prevalent form of energy storage on the grid was pumped hydro, which could provide hundreds of megawatts of backup power for balancing must-run coal and nuclear plants, and for demand management. But availability of pumped hydro was dictated by the location of these plants and building new hydropower plants was challenging because of environmental impact and NIMBY (Not In My Back Yard) issues.

There were a few examples, such as Uninterruptible Power Supply (UPS) systems that backed up critical loads such as datacenters that were rated at hundreds of kilowatts and used energy stored in lead acid batteries. DOE and EPRI experimented with several large rating energy storage systems using a variety of technologies, including lead acid, nickel-cadmium and liquid-metal batteries, superconducting magnetic energy storage (SMES), compressed air energy storage (CAES), flow batteries (Vanadium Redox), and concentrated-solar-thermal systems with molten salt for storage (California), all in the 2000–2015-time span. While each of these technologies had desirable attributes for high-power grid-scale applications, they did not address the fundamental question of market ramp up.

Utilities were not motivated to move these technologies beyond the pilot stage, and in any case could not move quickly enough to meet a start-up's cash-flow needs.

Manufacturers did not have alternate markets that could support rapid scale-up and a ride down the learning curve for such high-power systems. As a result, most of these technologies suffered from "death by pilot" and died on the vine. Without storage, utilities were forced to manage grid variability caused by increasing wind and PV solar penetration using slow market mechanisms, such as LMP (Locational Marginal Pricing), which did not address the fundamental issues of dynamic response and volatility and continued to cause problems on the grid. So even though technologies had been demonstrated, none of them seemed positioned to scale in a timely manner – maybe the process would take 50+ years, as had been the case for all new technologies up until now.

By now, we should be able to anticipate what comes next. While 2015 saw 221 MW of new storage come online in the USA, these systems were still mostly installed to meet a key reliability objective for the utility, or they were for demonstration projects to show key functionality and did not have to meet broader economic viability objectives (California and Hawaii). In our view, a transformative event in 2017 was the proposal by Tesla to use the batteries they were using for EV applications to support the grid, with a first application of a 100 MW 130 MWh system in South Australia. As opposed to a typical custom system that took many years to design, install, and commission, they delivered a system in 16 weeks from acceptance of offer – unprecedented speed of execution in the utility sector. They were able to do this because they modularized the design of the battery packs and inverters, building on and drawing on their experience with EV batteries. After all, the 130 MWh of storage represents less than the batteries used in 2000 cars, or half a day of production!

The ability to meet multiple market segment needs, fast and flexible action, and use modular distributed designs are key factors in the rapid growth that storage is seeing. It should be remembered that the EV industry is ramping up toward 125 million EVs in the USA by 2035 – representing more than 9500 GWh of stored energy and peak power handling capability of 25,000 GW (please recall US generation capacity is only 1000 GW today). On the other hand, NREL projects that the total grid requirement for storage will be 125 GW by 2050. While that is a sizable market, it is a small fraction of the potential EV market. Thus, in the near term, riding on the coat tails of the EV battery learning rate seems to be a good strategy.

As EV battery costs at scale approached $100/kWh in 2022, and a learning rate of 28% showed a possible path to $80/kWh or lower, it seems that the opportunities in the grid-storage market may explode going forward. We are beginning to see that an increasing number of utility-scale wind and PV solar systems are now being offered as hybrid systems, with integrated energy storage, and at a price point as low as $24/MWh for wind and $32/MWh for PV solar including 4 h of energy storage, both well below conventional coal and gas-based generation.

The energy storage makes the resources dispatchable and able to participate in the market, and to offer grid services to improve grid integration of renewables. Over time, other energy storage technologies will also take root, because they will be able to address specific market needs better. For instance, there is a strong need for Long Duration Energy Storage (LDES) that spans weeks or even longer. This is needed to address contingencies when the sun does not shine, or the wind does not blow for extended durations due extreme weather systems. Options being considered for this include green hydrogen, reflow batteries, and pumped hydro. However, as battery prices continue to decline and as their performance continues to improve, replacing them with alternate low-volume solutions that are only marginally technically superior may be very challenging.

The Changing Grid Edge

New Loads and Increased Electrification

We are getting a good feel for why rapid, almost uncontrolled growth was occurring in so many segments that we have traditionally associated with energy, especially in the areas of generation and transportation. But, as we have seen, the same exponential technologies were also driving a shift in consumer behavior, though now in ways that were meaningfully impacting the energy infrastructure, particularly the grid.

For instance, as video-streaming exploded (particularly during the recent Covid pandemic), the need to manage, transport, and deliver video data-streams across the globe also exploded, and increased by 700%. Edge datacenters sprouted like mushrooms, expanding to meet our voracious appetite for more and more bandwidth and to provide access to cloud-computing. Small edge datacenters consumed between 1 and 20 MW of power each, while larger datacenters could consume over 100 MW. About 10,000 new edge datacenters are being built in the USA, representing approximately 100 GW of peak power consumption, which the utilities have to provide. Video streaming also required rollout of advanced communications systems, such as 5G, which require a very dense last-mile signal delivery system. It is projected that 5G will consume roughly 2% of total electricity generated in the US.

We have discussed the many factors that drove the sudden emergence of EVs. All these new EVs meant less stops at the gas pump, as they would need to recharge their batteries somewhere else – i.e., on the grid. The presumed place was at home (remember EVs were originally considered to be a niche play for wealthy environmentalists). This issue has been explored since the 1990s by electric utilities when it seemed that EVs had arrived! The Infrastructure Working Council (IWC) has evolved standard charging rates, connector specifications, and communication protocols. Charging rates for home charging range from Level 1 at a power level of 1.5 kW at 120 volts AC for opportunity or emergency charging, to Level 2 at 240 volts AC and 30–50 Amperes (7.2–12 kW) for overnight charging.

EVs today are routinely realizing a range of 250 miles to 400 miles, with some manufacturers projecting an even higher range in the near future. These EVs are extremely energy efficient, realizing 3–4 miles per kWh, with total stored battery energy of 50–100 kWh. This shows that Level 1 charging adds only 4–6 miles of range per hour of charging, thus requiring many days to fully charge the battery – clearly not practical and meant mainly for emergency charging. Level 2 is better, adding 25–40 miles of range per hour of charging, getting to full charge during an overnight charging session of 8–12 h. As a result, most EVs integrate a Level 2 charger in the vehicle itself, allowing direct connection to the 240-volt AC circuit that is typically available in US homes. The maximum charging current that the EV draws can be programmed to not exceed the capacity of the plug and circuit. Level 2 charging also opens up the possibility of charging when the vehicle is parked, for instance, at work or at shopping malls, and to do so at a reasonable cost.

It was recognized early on that there would be an "infrequent" requirement where the EV owner would want to charge the battery rapidly so as to be able to continue on their journey with minimal interruption. The batteries, especially Li-Ion batteries, were capable of being charged at a high rate, provided the battery voltages and temperatures were managed properly. Because fast charging was expected to be only an occasional need, it was felt that EV owners would be willing to pay more for the convenience of not spending hours connected to a Level 2 charger.

To keep the cost of onboard hardware required for fast charging to a minimum, the IWC evolved the idea of Level 3 DC Fast Charging, where an external power converter was directly connected to the battery DC terminals through a connector and circuit breaker for protection. The vehicle would tell the Level 3 charger what level of current the battery needed at that time, and the charger would provide it. This has now evolved into a plethora of products that can charge EVs at rates from 25 kW to 350 kW and greater. As the batteries get charged, the vehicle can reduce the charging current as needed. Level 3 charging allows EVs to add as much as 60–80% range in 20–30 min (still a far cry from the 5-min refill time at a gas station, but much better than having to wait for 6–8 h while the battery is recharged). Between Level 1 and Level 3 charging, it seemed that much of the charging requirements of EVs were covered. But what happens as we begin to scale? The discussion until now assumes that the grid is available as a resource, whenever and wherever you need it, and that it has the capacity to grow to meet the needs coming from future growth of EVs. Is that true?

As EVs proliferate and achieve broader adoption, a few challenges are starting to become apparent. The cost of adding Level 2 charging capability in your garage can range from $1500 to $6000 if a service upgrade is not needed – posing a significant economic barrier for many households. Over 50% of US households (and an even higher percentage outside the USA) live in apartments or in houses that do not have individual residential garages. They often park their vehicles in communal lots or on the street. To provide ubiquitous Level 2 charging in apartment complexes will be very challenging. For instance, to provide Level 2 charging for 100 EVs would require a new service rated at 1.2 MW, in addition to the cost of installing 100 Level 2 charging portals, including the ability to bill for electricity delivered. There seems to be little motivation at this time for apartment complex owners to take on such an expensive facility upgrade. Table 4.1 provides a summary of the various EV charging protocols as it has evolved over the years.

The original hypothesis was that EVs would be secondary vehicles purchased by more affluent households. As EVs go mainstream and start demonstrating lower operating cost, they are attracting significant interest from lower income customers,

Table 4.1 Levels of Charging Requirements (Internal studies done by Center for Distributed Energy and Infrastructure Working Committee)

Charging requirements and issues

Type	Requirement	Charging issues
Level 1	120 volt AC 1.5 kW 4–6 mi/h	Only opportunity or emergency charging
Level 2	240 volt AC 6–10 kW 25 – 40 mi/h 6–10 h full charge	Require home/office garage 240 volt 30–50A service $1.5K–6K first cost
Level 3	DC Fast Charging 50–250 kW 150–1000 mi/h 10–20 min charge time	Charging station needed Visibility to availability Critical to ensure scaling $100–200K per port
Fleet	DC Fast Charging 50–250 kW	Depot charging for Van/Bus Overhead access for buses
eSemis	DC Fast Charging 250 kW – 1.5 MW 100–750 mi/h	Truck stops or service areas Complex electrical systems Challenging business model

and it is critical that equitable access is made available. Many US households, including low-income families, have multiple vehicles, many of which are used for work. All these vehicles would need charging if they were EVs, with many of them potentially charging at night once they get home from work. Further, the low cost of operating EVs is of great interest to low-income families, for people who use their vehicles as delivery vehicles, and for taxis and Uber or Lyft operators. The low operating cost is also of great interest for fleet operators, as the economic benefits accrue directly to the bottom line. Many fleets, such as used for package delivery and postal services, may have to charge multiple times in a day in a manner that is integrated with their operating schedules, often in locations that cannot always be specified. These EV users, attempting to complete a day's work, cannot afford to wait for 6–8 h while their vehicle is being charged.

Looking at the impact of EVs on the grid, we see a variety of factors becoming visible. For instance, in the USA, multiple homes are typically fed from a pole-top or pad-mount distribution transformer, often rated at about 25 kVA to 75 kVA. These transformers are not designed to operate at maximum peak power rating on a continuous basis (remember load diversity!). Rather, they are designed to heat up during a normal day's load profile, and to then cool down at night when the load is traditionally lighter. Now, if two or more neighbors purchase EVs (inspired by each other no doubt!), and then plug in their vehicles into their 10–12 kW Level 2 charger when they return home in the evening, we see that the transformer never has a chance to cool down. This can shorten the average life of the transformer from 30+ years to less than 3 years! Utilities do not monitor distribution transformers, much less track which ones have EVs on them and are running hot. As EVs scale, this clearly would require a substantial and expensive distribution infrastructure upgrade.

It is becoming increasingly clear that the availability of ubiquitous charging infrastructure is a critical requirement for the broad adoption of EVs and for continued growth along the trajectories that are projected. One can also see that the grid will be increasingly stressed as EV loads grow – though this fact is not well recognized. A lot of the analysis that has been done to look at the adequacy of the grid to meet EV growth projections has been based on the amount of energy required. Using that metric, if the EV fleet grows to 125 million EVs by 2040, the fleet will consume ~125 TWh/year – or roughly 3% of the electricity produced in the USA today, representing a miniscule 0.15% annual growth rate that is very manageable over a 20-year period, and should not be a concern. However, there are several areas of concern that are already becoming visible and need to be addressed.

A key factor that the above discussion highlights is that most EV owners and operators may not be able to access or rely on Level 2 charging *and will frequently need to use fast charging*. In many cases, especially for low-income customers, fast charging may be the only practical means of charging their vehicles (even more relevant when the EV is a shared vehicle). This is quite different from the original scenario, where fast chargers, considered a luxury, were to be infrequently used. The cost of a DC fast charger rated at 250–350 kW is in the range of $50,000 to $100,000, not including the fast-charging station itself. A facility with six charging stations would need an electrical service sized to provide 1.2–2.1 MW of charging power (which is very expensive when compared with a 70-kW service for six Level 2 chargers). Grid infrastructure has to be sized to deliver the peak power needed by the load(s). This includes generation, wiring, transformers, protection, and metering. As such the cost of providing a grid connection is directly related to the peak power that will be drawn. Until EV volumes build up and a deployed fast charger is used for a significant part of the day, we will see poor utilization of the charging facility (as an individual vehicle would only charge for 20–30 min). In many ways this is a double whammy – because the energy revenue derived by the station is for only one vehicle (say 50 kWh), but the cost of the expensive charger and the facility to supply peak electrical demand has to be covered by the business model for the station.

Traditional electricity billing models that charge only for energy delivered do not adequately cover the cost to the utility of providing the service for such high-peak loads. As a result, utilities typically levy a "demand" charge on top of the energy cost itself. EVs using fast charging have to cover both demand and energy costs plus cost-recovery associated with the amortization of the charging station itself. It is clear that, especially in the early years when EV population is growing slowly, the cost of fast charging under today's business models will be very high.

Yet, unless a nation-wide network of public-access Level 2 and Level 3 chargers is rolled out and available across the entire country (including rural and peri-urban areas), issues of range anxiety will not be addressed, and EV growth rates will likely slow down. In recognition of this, several companies, including Tesla, Electrify America, and Charge Point, have been rolling out hundreds of thousands of fast charging portals. The Biden administration has also jumped into the fray and has committed $5Billion to deployment of charging infrastructure for EVs, with an overall goal of 500,000 charging points across the country. For a mid-level target of 125 million EVs in 2040, the number of chargers needed is probably closer to six million Level 3 fast chargers, in addition to ~30 million Level 2 chargers. It appears that we are far from that target and that the issue of charging infrastructure continues to be a major one and could become an impediment to the rollout and scaling of EVs.

One other key element that is not being considered as EV charger companies look at rapid deployment of fast charging infrastructure is the interaction of fast chargers with the grid, particularly as EVs achieve scale. If we assume that the EV population grows by 2040 to the mid-level estimate of 125 million EVs, and if we assume that only 8% (ten million) of the EV population were to connect to the grid at a particular time (a fairly high probability occurrence) using a fast charger that delivered 100 kW (remember fast chargers today are rated at 50–350 kW), we would need additional peak generation of 1000 GW to serve this new coincident peak load. Given that the total US generation capacity today is 1000 GW, it is difficult to see any scenario under which we could successfully double traditional generation to serve this new load in a short span of 18 years. And who would bear the cost of the new infrastructure needed to serve this poor capacity-factor load?

This demonstrates that the interaction of the EV charging infrastructure with the grid is a key and often ignored issue, whether it is addressing EV charging points in rural communities, or the question of deploying 500,000+ fast-charging stations in the near future. It also highlights the fact that while the EV's key metric is the energy delivered to the battery, we see that the cost of delivering that energy is inextricably tied to the peak charging rate and can have a severe impact on the cost of driving an EV, especially under current charging-station business models. The additional need to preemptively rollout a ubiquitous and publicly accessible charging infrastructure before the population of EV is actually there to use it puts tremendous strain on the charging-station business model and could raise the cost of driving EVs to levels that challenge their economic viability. Given the oversized impact that the grid has on EV charging, it is surprising that most EV charger manufacturers do not even mention the grid and the impact that it could have on their business.

While the above discussion has focused mainly on the interactions between EV charging and the grid, we see many other trends becoming visible that could see tremendous acceleration over the next few years. The operating cost benefits for electrification of transportation accrue not only to EVs, but also to vans and trucks, including the larger Class 8 vehicles, often referred to as e-semis. The operating cost of a large electric truck can reduce from $0.52/mile for diesel trucks (even more as diesel prices increase) to <$0.22/mile (~60% reduction). This potential for cost reduction, and not the climate impact, is driving major truck manufacturers to release new models for electric trucks and trailers. Similarly, electric buses are already making significant inroads and are seeing increasing deployment.

The larger vehicles have larger battery packs and higher power inverters and motors, and also need to be charged at higher power levels, with projections of 500 kW to 1500 kW per vehicle. The anticipated availability of these vehicles in the market is estimated to be over the next 3–5 years. A typical truck-stop or warehouse may simultaneously be servicing 50+ trucks, suggesting that the grid has to provide 50–75 MW of power to this facility, a facility which today probably has an electrical service of <100 kW. This represents a major investment for the utility and would typically require the build of a new substation and possibly new transmission lines – a process that could take 10+ years. The above discussion of peak versus average load and infrastructure utilization would again influence the actual cost realized by fleet owners. It should be noted that truck charging would add even more to the peak charging requirement on the grid that was identified above.

Similarly, driven by factors such as need for long duration energy storage, plummeting prices of renewable energy, and reactions to climate change by investors and corporate sustainability goals, there is new interest in technologies for generation of green hydrogen, electric aviation, water purification, direct air capture, and electrification of traditional industrial processes. In addition, many cities are putting a moratorium on new natural gas service, and are driving adoption of electrical technologies for heating, cooling, and cooking. All these advances are in very early stages of development and demonstration, and as such are not yet impacting the grid. However, if these technologies were to scale rapidly over the next 5–10 years (as seems likely), we would see similar issues emerge in terms of grid integration as we have seen above. Again, it should be noted that all these developments are occurring outside the sphere of utility influence and are progressing at a pace that is much faster than the integrated resource planning cycles for most utilities.

New Generation Resources

Not only is the grid-edge changing due to new loads, but also because of new distributed energy resources that are being deployed at the distribution level. These DERs (Distributed Energy Resources) include PV solar, batteries, and microgrids – addressing issues of cost, reliability, and resiliency. We have already discussed the impact that Feed-In-Tariff (FIT) and net-metering had on residential rooftop PV solar in Germany and California. As PV prices declined, the demand for both residential and commercial scale PV solar continued to rise in the USA, reaching cumulative installed capacities of 19 GW and 15 GW, respectively, by 2020 (not including an additional 59 GW of utility-scale PV solar). This 34 GW of distributed generation was all located at the grid-edge, with an additional 15–20 GW of new capacity being installed every year. And, because the deployment of DERs was not uniformly spread across the states (a result of policy and incentives), the actual penetration of renewables on some distribution feeders in specific states (e.g., California and Hawaii) could exceed 100% at some times, causing reverse power flows and a host of issues for utilities in terms of grid integration of these DERs.

The other area seeing rapid growth was energy storage. Lower battery prices saw the emergence of new products such as the Tesla Powerwall, which could improve the resiliency of a home or business. True, these were still very expensive, required a lot of customization, and were only affordable for the wealthy or for industrial users (e.g., Walmart) who wanted to be green even as they became more resilient. Given rising concerns of resiliency and climate change, many critical facilities (e.g., datacenters, army bases, hospitals, and airports) and some commercial customers started looking at backup power and microgrids to power their critical loads through unforeseen long-duration emergencies. This was achieved with fuel cells, microturbines, or with old-fashioned gas fired generators, representing more than 100 GW of potential distributed generation at the grid-edge. All these changes were occurring on the distribution network at the grid-edge, a place where utilities have poor visibility and even poorer control. As penetration of distributed resources on the networks increased, especially in places like Europe, Hawaii, and California, we increasingly saw the grid challenged to manage the integration of these new resources with traditional grid operations.

The period from 2000 to 2020 also recorded a dramatic increase in grid resiliency events, many of which, including their frequency and devastating impact on society, could now be better correlated with climate change. High-impact low-frequency (HILF) events caused by hurricanes, wildfires, flooding, and extreme cold snaps (e.g., polar vortex) – events that not only impacted the distribution networks and the grid-edge, but also impacted the bulk power system, including generation and transmission.

Extended outages, ranging from weeks to months, in diverse places including Puerto Rico, New York, Texas, Louisiana, and California, caused death and devastation at unprecedented scale and underscored the widespread nature of the problem, as well as the difficulty that the utilities had dealing with disruption at this scale. It called into question a paradigm that took pride in a reliability level of 99.99% on the bulk power system but could not deliver on that promise when faced with major disruptions, causing millions to suffer the consequences of losing what we have taken as a fundamental right to energy access in the USA. In Puerto Rico, restoration of power for some people took as long as 240 days, a long time to live without power in the USA in the twenty-first century. This has brought into sharp focus the issue of resiliency of the power system and the related issue of energy equity. These events did not occur in an emerging economy country where they were still struggling with establishing the grid, but in the very country that had defined what universal access means almost a hundred years back.

It is important to understand that utility resiliency processes are largely geared toward restoration of the distribution system and not the bulk power system itself. The bulk power generation and transmission system require specialized and highly customized components, such as high-power-transformers that can take 12–36 months to build – a long time for the grid to be down. Reliability on the bulk power system is typically achieved by building in redundancy in generators, transmission lines, and substations, which can be compromised when a wide-ranging geographical area covering many critical assets is impacted. In such instances, service can take a long time to restore, often requiring emergency response with support from many neighboring utilities.

Maintaining critical power during outages following HILF events thus involves both the bulk power system and the distribution network. Power needs to be restored not only for load centers and services that utilities deem critical, but for all people – each of whom considers their own assets to be of value. The inability to restore power, often for weeks and months, is becoming a major issue. The top-down architecture and design of the current power system does not lend itself to maintaining power at the grid-edge when the bulk system is down. Black-start and power restoration today is a highly complex process that can take weeks. Backup power solutions, such as backup generators, whole-house UPSs, and islanded microgrids, are not equitable because they are highly customized, too expensive, and cannot easily be integrated with normal grid operations and services for the broader community. It is a tragedy that even as more and more DERs are being deployed, such as rooftop solar, energy storage in EVs, backup power for data centers, and critical loads, utility customers (who are all at the grid-edge) remain vulnerable to power outages and can experience poor reliability and resiliency.

Clearly, those who are wealthy enough to afford backup power for their businesses and homes do not stay in the dark, underscoring the highly disproportionate and inequitable impact that climate change is having on the poor. It was only adding insult to injury that even those homeowners, for instance in California, who had PV panels on their roof and EVs in the garage still had to sit for weeks in the dark as they coped with massive forest fires. Not only was this a failure of imagination, integration, and technology, but also reflected poorly on the promise that utilities had made; that a top-down centralized and regulated grid paradigm could deliver on all the energy needs of modern society. By way of example, "Finding 5.4 in the NASEM Report on the Future of Electric Power in the US" contemplates a future grid where reliability and resiliency are achieved from the grid edge, and access to low-cost power could be provided by the bulk power system, when it is available (which is 99% of the time) [7]. We are a long way from that goal, but it underlines the need for smarter generation resources at the grid-edge, resources that can provide resiliency under HILF events and can do so for everyone without blowing the budget.

We do see a lot of distributed generation already being deployed at the grid-edge, at datacenters, for microgrids, and to access low-cost solar energy on residential and commercial rooftops. We see a lot of energy storage, in particular batteries, being deployed at the edge, sometimes to meet residential or business needs, sometimes to support the grid, and in the form of electric vehicles that each carry large batteries and are often connected to the grid. We see battery charging stations coupled with batteries and PV solar canopies that can reduce the peak demand for the facility. Yet, all these resources are single purpose devices, deployed in the service of one owner for one purpose, decoupled from the rest of the system, and unable to provide support or services to neighbors and to a grid in need – that is, these resources are limited by regulations, technology, and business model. If projections on the growth of distributed energy in all the above sectors are summed up, it far exceeds the total generation capacity in the USA today – and it is all located at the grid edge. *This presents perhaps the biggest opportunity for a major paradigm shift in the way we think about the grid – a move toward a more distributed and decentralized system, where private and regulated owners of resources collaborate to support the overall requirements of the new grid.* However, the utility sector is far from believing that such a scenario could evolve, or that they need to change their operating paradigm so dramatically.

The Grid Is Changed: Forever!

All the changes we have described – in renewables, storage, EVs, and at the grid-edge – are changes that have occurred fast and outside the sphere of utility control. Yet, these are all directly connected to the grid and exchange energy with it in ways that were not anticipated by the utilities. This has resulted in significant change in the physical behavior and response of existing grid elements – even though utilities did not want such change to occur. It is vital to understand what these changes are, what the impact on the grid will be, and how the impact will be further exacerbated as the many driving factors continue on their path of unfettered and uncontrolled growth.

Renewables (now controlled by power electronic inverters and not synchronous generators) introduced something new onto the grid, something that was not understood or anticipated. As wind and solar were increasingly deployed, first in Germany and Europe, and eventually in the USA in states such as California, a new host of issues started becoming visible. We have seen that utility-scale wind and PV solar farms tipping the scale at 300–1000 MW and more were sprouting like weeds, with a need to bring this power to the grid. This required new transmission, which in the USA has historically and culturally been challenging (remember NIMBY!). Even if transmission could be built to connect the wind or PV farm to the grid, controlling power flows on the meshed grid was a challenge and created congestion and reduced system utilization.

In China and India, a few HVDC and multiterminal HVDC (MTDC) systems were built to bring the power close to load centers and to inject this new power into the grid at optimal locations. This in turn raised the question of what happens when a HVDC line is lost due to a fault, reducing overall system capacity, and potentially causing system instability. Power throughput from the renewable resources had to then be throttled back to ensure that under the contingency of a lost HVDC line, the system did not become unstable. Point is, these are all very complex system integration issues, and moving fast with very large investments can result in undesirable and unanticipated outcomes that can seriously jeopardize future adoption rate as well as the return on the investments.

What is needed is a more flexible and adaptable approach: a way of making no-regret investments that can allow pivots or course corrections as new information becomes available, or as new technology advances. This is fundamentally different from the traditional method by which grid-related investments are made today and also requires a new way of operating the grid that often contradicts current practices.

With the new and variable renewable energy now injected on the grid, there was also a need to dynamically balance the mismatch between generation and load – and market mechanisms were too slow and relied on communications, which could have latencies or be interrupted. Further, the distribution networks were not designed for reverse power flows that could now occur, or to manage the variability of AC voltage that was caused by the PV panels and inverters. Much more important was the fact that these early inverters were regarded by the utilities as a nuisance. As a result, these inverters were required to "trip" when they sensed abnormal conditions on the grid, such as deep voltage sags or under/over frequency events, or when the grid was not present (a condition referred to as "islanding"). As these issues were repeatedly observed on the grid, standards were then developed or modified to ensure that all grid-connected inverters complied with these "grid-codes." However, these standards were developed only after the issues were observed in the field, and then went through long consensus and subsequent product development cycles. This also suggested that these standards were invariably 8–10 years behind the fast-moving technology cycles that inverters were following. As inverter deployment grew to hundreds of gigawatts per year, new issues started becoming visible, which were not covered even by the latest standards.

For instance, with a rapidly growing base of PV generation, Germany's grid was seeing very high inverter penetration levels and was particularly susceptible to grid–inverter interactions. They almost had a large-scale blackout in 2008 when all the small rooftop PV inverters tripped (as they were required per existing grid-codes) on a system fault that caused a deviation in the line frequency – a "frequency excursion." This event eventually resulted in new requirements for inverters that now were not allowed to trip but had to ride through the disturbance and keep operating under grid fault conditions.

The seriousness of this issue required reprogramming or replacement of 315,000 PV inverters that had already been installed in Germany – a very expensive exercise. This also highlighted the difficulty of communicating with and coordinating the actions of a large number of devices in the field that were individually small rated but collectively had a big impact on the grid. These distributed inverters could respond autonomously and very quickly (in milliseconds) to local disturbances, but as a result were also blind to what the other inverters were doing and had a tendency to interact with each other. Academic exercises, requiring perfect knowledge of a neighbor's control algorithms and the ability to precisely compute what each inverter should do and to communicate that to neighboring inverters were impractical. It was also unreasonable to expect the utility to know the conditions at each inverter location in real time, and to be able to compute and "dispatch" a command (as it could do with 5–15 min latency for the smaller group of large-rated conventional generators). Even if such centralized but distributed control could be done (given so many different inverter owners and vendors from many different countries), it would be unwieldy, unreliable, very expensive, and challenging to scale.

This increasing population of distributed fast-acting control assets is a major challenge with a future decentralized grid that still has to be solved.

Another concern that was slowly becoming visible was the change in the basic behavior of the grid to disturbances and transients. The traditional grid was supported by thousands of very large spinning generators, with massive rotating inertia like an enormous, distributed flywheel. These machines were all electrically (and thus mechanically) linked together and intrinsically gave the system ride-through capability under disturbances, as well as provided a high level of short circuit current capability to trip passive devices, such as circuit breakers or fuses, and to thus remove faulted sections of the grid allowing the rest of the system to keep operating. The transient and fault mode behavior and operation of the grid has traditionally been governed by the properties of such protection devices. As millions of inverters replace rotating machines (already happening in Europe and in parts of the USA), there is a substantial loss of rotational inertia on the system, as well as a dramatic reduction in fault-current sourcing capability. This results in a significant change in system response to large faults, often causing large transient frequency excursions (called Rate of Change of Frequency or ROCOF events) when disturbances occur, with the potential of tripping existing protection devices in the system or causing instability. This is also becoming a major area of concern for grid operators.

The growth in the number of inverters had another profound effect that was also not on the radar of grid operators and electric utilities. Inverters, including those made today, are generally meant to push power into the grid. They assume that the grid voltage is present and that the inverters operate in "grid following" mode to push real power back into the grid as needed. The more advanced inverters can also provide "reactive power" support to maintain voltage profile, manage phase unbalances, and suppress undesired harmonic frequencies. Energy storage systems can also pull power from the grid as needed. However, as inverters become a larger part of the system, they also have to play a role in "forming" the grid, because rotating generators may be few, or in the future, may not be there at all. Today, inverters do form the grid in some restricted applications – such as in UPS systems and microgrids, but only with very detailed knowledge of the system and tight control over all the elements – essentially as a small-scale grid operator. When applied to the broader decentralized grid, this level of detailed accurate information is not there. How can we then control such a system, especially with the high level of safety and reliability that we take for granted today?

Further, in grid-connected inverters, how does one know that there are, or conversely are not, any generators on the system, and whether the inverters need to be operating in "grid-following" or "grid-forming" mode. One cannot rely on communications to realize control at the millisecond level, as there may be latencies. *Unlike our phones, the grid must run in real time, even when data packets are lost!* So how do we control inverters so that they collaborate with each other and do not fight each other or cause instability. While the "steady-state" operation of such systems is well understood, the ability to control the very-fast-response inverters so they operate without low-latency communications, and can support the grid in all modes, including under transient and fault conditions, is a challenge and still a work in progress.

This represents another major transition for the grid – from a centralized system with hundreds of tightly controlled generation assets per control area to a highly distributed and decentralized system with millions of generators, batteries and loads that need to be coordinated and controlled, often with millisecond precision – a problem worse than herding cats!

The above discussion has also brought into focus the need for dynamic balancing of load and generation on the grid. The traditional grid operates with a "generation follows load" paradigm, typically realized with a significant surplus of generation over demand. With high penetration of variable resources, the need for dynamic balancing only increases. This can be achieved using energy storage, where energy is stored in times of surplus, and then used when needed. Storage is needed for different durations – short (seconds to minutes), medium (hours to days), and long (weeks to months). The storage could be needed at utility scale on the transmission or distribution network, or at local/residential level at the grid-edge. The utilities have used pumped hydro (ratings of up to 1.2 GW) for decades, especially for balancing the must-run nature of nuclear plants.

As penetration of renewables has increased in Germany and Denmark, there has been a need to balance the variability with the demand. This has been done by building a bidirectional HVDC link between the European mainland and Norway, which supplies excess renewable power to Norway during times of surplus and pulls hydropower (also renewable) from Norway when there is a shortage on the mainland. This approach does not require pumped hydro and only uses existing hydro plants by controlling them differently. This is a great example of how existing hydropower plants can be harnessed to dramatically increase renewable penetration on the grid in a cost-effective way. Since there is not enough storage on the US grid, dynamic balancing is a major concern, and this approach could provide a viable strategy.

Even as these massive disruptions (DERs, EVs, and resiliency) began unfolding at an accelerating pace (especially after 2010), most US utilities typically remained focused on the issue that was traditionally on their plate: power system reliability, with an underlying assumption that if bulk power was available, then high reliability could also be ensured at the distribution level to serve customers who were at the grid-edge. Utilities had also slowly started thinking about enhancing distribution system reliability to improve SAIDI (System Average Interruption Duration Index) metrics, installing reclosers on the distribution system to rapidly isolate faulted sections and to reconfigure the system to restore power to as many people as possible. The AMI (Automated Metering Infrastructure) system, which was a key element of the Smart Grid, continued a slow rollout but proved to be very challenging and expensive to operationalize, especially to extract the full benefits of visibility and control at the grid-edge. As a result, for most utilities (including many smaller coops), the grid-edge still remained dark with poor visibility, and the promise of advanced system awareness coming from AMI remained only a promise.

Digitalization has also created its own set of new problems for the grid. The open internet has dramatically multiplied problems with cybersecurity, cyberattacks, and the vulnerability of operating systems – threats that are both financially and politically motivated. The bigger, more centralized, and more critical the operation, greater is the threat and impact of a cyberattack. We continue to see examples of such threats, at national and global scale, on an almost daily basis. The grid, especially a centralized grid, presents a massive target for cyberattack, especially from hostile state actors and other malevolent people. As millions of intelligent grid-edge devices, such as Nest thermostats, EV chargers, rooftop solar and battery energy storage, get deployed, the chance that a major coordinated cyberattack on aggregated populations of such devices will succeed, only grows. And because all these devices also exchange energy with the grid, the possibility that the grid will be impacted through a distributed but coordinated cyberattack at the grid-edge is also increasing.

The changes wrought by the disruptions we have covered in this section were systematically resisted by a large part of the electricity industry (and regulators) in the USA, citing concerns of reliability, grid integration, protection, coordination, poor asset utilization, stranded assets, and higher cost of energy. This was borne out for many utilities that had to manage high DER penetration levels, where customers often paid more for energy than in other areas where DER penetration was low. As recently as 2010, most utilities did not believe that renewable energy and storage could ever provide a cost-competitive option to dispatchable generation – the heart of the electric grid. They could not conceptualize that the steep learning rates for PV, wind, storage, and EVs, as well as digitalization in general, would actually translate into a fast-moving tsunami that could wash away a hundred years of practice and dogma that was taken as gospel.

The last 20 years has proven to be a challenging time for the electricity industry. It was not like the days of Edison, as if utilities had a greenfield situation to build the grid that we needed, from the ground up. The industry had to keep operating, meeting today's demands for availability, reliability, and low-cost of electricity, while at the same time transforming into a green and carbon neutral business. As stated by Mike Howard (now retired CEO of EPRI), *"what the industry had to do was akin to flying an airplane and swapping out all the parts to transform it completely, all while it was still flying!"* [8]. This challenge also meant that trillions of dollars of assets – making up the still-performing grid – would soon be substantially devalued.

As stewards of their assets, utilities were answerable to their investors – who were in turn looking for undiminished returns and continued growth. It is not surprising that even with the best minds focused on the issues, they consistently missed predicting the disruptive future that was emerging and the actions they needed to take to manage the impact, until it was very late (almost 2020!) [9]. At the 2021 EPRI Summer Meeting, EPRI leadership and a large part of the US electricity industry leadership finally proposed targets of 50% decarbonization by 2030, and 100% by 2050. While utility leadership seems to finally understand the necessary end point, without an assured pathway that is technically and economically viable, achieving these goals seems at best aspirational. It is interesting to note the parallels with the fossil fuel industry, which also saw the impact of disruption caused by both the transportation and electricity generation sectors, and where leadership also dragged its feet and obstructed change for as long as possible – but is now faced with a new reality!

At the 2018 IEEE Power and Energy Society General Meeting in Atlanta, the challenge posed by high-inverter-penetration DER-rich grids was very clear. In a panel session with more than 2000 power system engineers in attendance, all of them experts on systems level issues and grid operations, there was near unanimity that all major issues with scaling up to a grid dominated by renewables were essentially solved! At the same time, in another panel session with 50 trained power electronics engineers, the people who actually designed these inverters, the verdict was that we did not "have a clue" as to how to solve these problems [10].

Similarly, in a panel session organized in late 2021 by the NASEM Board for Energy and Environmental Systems (of which Divan is a member) on the controllability of the future grid, we had the CEO and CTO of one of the largest independent system operators in the USA, as well as technologists working on grid-connected inverters, agreeing that the grid was getting increasingly challenging to control as the penetration of inverters on the system is increasing. However, this has not stopped us from deploying 100 GW/year in new grid-connected inverters! But there is hope that we will have a solution by the time we need it.

The above discussion also underscores another major evolution with DERs, that is, a slow migration to a more distributed (and decentralized) architecture for the future grid, as PV, wind, and storage are scattered around the grid, both on the bulk system and the distribution system. Are we ready for it? Do we understand the challenges that a massively decentralized and distributed future grid will pose? Why would we want to do this in any case? Isn't a centralized grid much better?

Why did so many of the energy industry experts miss the mark by so much? Even the recently released report by the National Academy of Sciences Engineering and Medicine – "The Future of Electric Power in the United States," acknowledged that projections and the underlying assumptions by many leading organizations, including the EIA, WEO, and IEA, were all consistently wrong over a period of two decades. This may only have been of academic importance if global investments and strategies were not based on these projections. Unless we understand why such gross and continuing misjudgments occurred, we are probably doomed to repeat our mistakes (as we have done for the last 20 years!). But before we can address what needs to be done, we need to first look at some additional contributing factors.

Other Contributing Factors

This discussion would not be complete without an examination of the role that China has played over the last 20 years in driving this disruption. The last two decades have seen a substantial shift in financial and technological power as China, starting essentially from scratch in 2000, grew to the second largest economy in the world, lifting millions out of poverty. China started to change course from its hard core communist economic policies in the early 70 s, and under Deng-Xi Ping in 1976–1977 began to open China for low-cost consumer manufacturing. Wall Street

& Corporate America exploited this by squeezing out additional profit margins through outsourcing manufacturing to lower wage countries. This also allowed them to shutter expensive manufacturing plants, layoff expensive workers, and reduce longer term R&D initiatives.

The rise of Asia's commodity manufacturing and increasing technology maturity perhaps put the biggest pressure on traditional innovation centers – the market leaders' research labs, where sustained efforts over decades had been required to de-risk new technologies and to ensure a smooth transition so that their existing investments in older technologies were not jeopardized. These competitive pressures resulted in a dramatic shrinking of the core long-term industrial R&D capability in US companies, with a strong shift instead toward near-term development and demonstration initiatives that could be brought to market in 2–3 years. The inability to maintain a competitive edge in manufacturing saw a dramatic shift in manufacturing to Asia, in particular China.

As discussed earlier, China forced vendors interested in supplying Chinese markets to share their technology, and then capitalized on this technology transfer, rapidly leveraging it to build internal R&D capability to advance this further. Their nationally specified priorities, low costs, captive universities, and nationalized R&D process allowed them to focus on rapidly moving core-technology elements and to move them forward. A large cadre of entrepreneurs (many of whom were from universities such as Tsinghua) were willing to use government support to develop solutions. Chinese willingness to have large, nationalized utilities commit to developing and testing these new unproven solutions on their system, and to then rapidly deploying these high-risk solutions at scale even as the solutions continued to be improved, provided a unique competitive advantage that vendors in the West simply did not have.

University research centers were established in China with a focus on PV, wind, power systems, batteries, electric transportation, and other critical areas to provide design and testing support for companies developing these technologies. Even as Chinese manufacturers were driving down costs, the internal market for PV and wind opened up, as the government committed to spectacular deployment targets of 100 s of GW. This also resulted in a strong need for building HVDC lines to interconnect the generating regions with the load centers, to become pioneers in multi-terminal DC (MTDC) systems, and to strengthen the AC grid to accommodate growing load demand, as well as this new renewable resource. This created brutal competition between Chinese vendor companies, squeezing inefficiencies out of the process, and focusing the outcomes more on fundamental properties of the underlying technologies. The companies that survived this internal winnowing were extremely strong in terms of technology, experience, and cost. They were also nimble, able to respond to fast-moving technologies while simultaneously operating to keep prices down. This gave them immense competitive advantage globally, as they began to increasingly dominate their respective fields.

The USA, as we have seen, operates very differently. For instance, one important factor to consider are the players in the utility sector. Electric utilities and grid operators:

- Are encumbered by federal, state, and local regulations.
- Report to public utility commissioners who are political appointees and typically may not deeply understand the challenges or opportunities that this technology-driven disruption brings.
- Operate with a utility workforce that does not understand the new technologies and lacks experience in how to operate in a high-velocity, high-risk environment.

Given these constraints, it is not surprising that the grid represents perhaps the biggest challenge in this energy transition. The fact that the electricity industry spends less than 0.1% of revenues on R&D, extremely low for an industry that is in the midst of massive disruption, raises further questions about its ability to manage and drive the transition to a soft landing such that its businesses are fully protected. It should be noted that technology-driven businesses (which electric utilities now are) typically spend as much as 10% or more of revenues on R&D, 100× more than the electricity sector! [11]. If utilities and grid operators cannot (or are not allowed to) innovate and adapt quickly, the price of local generation could fall to the point where more utility customers would consider grid defection, initiating a fragmentation and possible death spiral of the electric utility industry. This would create chaos along with massive financial impact and could easily derail the energy transition. A path forward is needed to achieve the energy transition, one that is not disruptive to our daily lives and our economic well-being.

Key Takeaways

In this section, we have taken a look "under the hood," trying to understand the major drivers for this very wide-ranging global energy transition that seems to be underway, one which seems to have caught everyone by surprise. Given the very complex and interconnected set of issues – including science, technology, innovation, entrepreneurship, manufacturing, supply-chain, policy, economics, finance, geopolitics, and consumer behavior across a range of sectors, and given the high velocity of the change across this entire interconnected space – it is not surprising that we have a very noisy environment where it is challenging to extract the key factors that are driving the process forward. Surely, there are major gaps in the way we ourselves have looked at the issues, but we feel there are a few key factors that are at the heart of why and how this fast change is occurring, factors that have not received as much attention as they should have. These factors, we feel, can also provide key insights into how the next 20 years and beyond are likely to unfold.

First, at the heart of this major disruption are unrelenting and unstoppable advances in fundamental science and a set of related technologies that now allow us to understand, predict, and manipulate materials at the micro, nano, and almost atomic level. Much of this is based on an improved understanding of quantum mechanics, realizing functionality that was simply not possible with classical physics and chemistry based on observations at the macro-level. This has led of course to new energy sources, such as nuclear energy, but has also opened the door to semiconductors and a tsunami of continuing advances in digitalization, communications, computing, power processing, Photovoltaics, Open AI (Artificial Intelligence), and Quantum Computing. It has led to new experimental and analytical tools, such as electron microscopes, atomic surface microscope, and X-ray lithography, which have allowed the design of new nanomaterials and complex microscopic structures that do not exist in nature, thus providing visibility as well as the ability to build devices and structures at the nanometer and atomic level.

It seems almost counterintuitive that the above discussion on energy and its disruption does not even mention primary energy sources. For experts and users steeped in the traditional processes of the extraction, processing, transporting, combusting, and utilization of bulk fossil fuels to realize the energy-based services that we need, this almost feels like it is a science experiment done by people who do not have a real and practical understanding of how massive and expensive the existing infrastructure is, infrastructure that was built over a hundred years. It seems almost inconceivable that the new technologies can achieve scale rather quickly – a process that history suggests should take another hundred years!

But they have been proven to be wrong, again and again, and in almost every sector. We did not have this disconnect in computation or communications because there was really no other prior way of getting the new functionality. Intel did not have to dethrone an incumbent abacus manufacturer whose products were being used to build mechanical computers – such machines simply did not exist (other than as lab curiosities). However, in the field of energy, because society already uses hundreds of quads of energy every year, there is a massive base of industries and stakeholders, people who have the financial resources, political clout, and all the motivation in the world to make sure their businesses and business models are not disrupted by these upstarts [12]. And they may still succeed, or at least delay the transition. However, the exponential growth trajectories that we are seeing for many of these technologies point to a very likely new horizon where the disruptors will prevail.

Second, we believe that another dominant underlying factor common to all these unexpected growth stories is a steep and sustained learning curve that drives a key, value-based, technology-agnostic metric to assure continued growth and expansion of the industry sector. For the right solution, there is an addressable global market that increases dramatically as price drops (often accompanied with improved performance). This potential for exponential growth in volume drives rapid price declines and fuels further market expansion, attracting new entrants with technologies that are novel and competitive. Our deep human desire to explore, create, innovate, and compete allows a global pool of researchers, technologists, and industry leaders to move forward using a multiplicity of technologies and approaches. A concurrent and rapid expansion of the knowledge base continues to create new and better approaches to solving the key problems, which keeps the virtuous cycle going.

In addition to the advances in materials and our understanding of their use at the micro, nano, and atomic levels, we believe, another dominant underlying common factor to all these unexpected and exponential growth stories is a steep and sustained learning curve that drives a key value-based and technology-agnostic metric to assure continued growth and expansion of the industry. This allows and motivates a global pool of researchers, technologists, and industry leaders to rapidly evolve the technologies using a multiplicity of approaches and cross-technology synergies.

We see this story being repeated in integrated circuits, memory, photonics, electronics, communications, AI – as well as in energy-related technologies such as photovoltaics, power conversion, batteries, and electric vehicles. Unlike traditional technologies that evolve slowly, the convergence of fast technology cycles, continuously improving functionality and performance of these exponential technologies, all achieved even as costs decline, also creates opportunities to refresh the deployed base of products and to avoid market saturation, with several product examples validating this consumer penetration and growth trend, e.g., mobile phones, televisions, computing, cars, and other devices. There are many more emerging technologies that also seem to demonstrate similar steep and sustained learning curves, technologies that have the potential for massive global impact. These rapidly emerging technologies – which we will examine more closely in Chap. 8 – include green hydrogen, Direct CO_2 capture from air and its permanent sequestration, water purification, electro-metallurgy, cement production with no CO_2 emissions, solar to fuel conversion, and many others.

A third key factor is a transformation from centralized to distributed systems. If we look at twentieth century technologies, where we think of technology as evolving relatively slowly and adoption rates as tightly controlled, we see that scaling is generally addressed through incremental innovation and by exploiting cost reductions that come from economies of scale. This leads to very large plants and centralized infrastructure. Large steel plants, oil refineries, factories, pipelines, as well as utility-scale nuclear, hydro and coal power generation with nationwide transmission infrastructure are examples of this. These are generally large custom-designed systems with complex economic models, where the scoping, permitting, building, and commissioning can take many years, with decades of operation under a service agreement that allows economic recovery over the life of the plant, under an assumption that prices will generally remain stable. Plant Vogtle in Georgia, the only nuclear plant being built in the USA today, with a rating of 2500 MW, has been under construction since 2013, has an expected (and still escalating) price tag of about $34 billion, and symbolizes the challenges. Similarly, with the sharp drop in natural gas prices that came from fracking, we also saw a spike in the deployment of 500 MW natural gas generation at $50/MWHr replacing coal plants in the 2015 timeframe, deployments that are already seeing an unviable financial model.

We saw these large plants being challenged by PV solar and battery energy storage, technologies where prices declined by more than 100× over 20 years. Over the same period, the products went through many technology cycles, advancing in performance and functionality even as price declined. Even large PV solar plants and wind turbines that were built 10 years ago were suddenly uncompetitive when compared with more recent deployments. The ability to accommodate fast technology change was achieved through the use of more distributed and modular solutions that were interoperable with previous generations of already deployed technology. An example is the ability to use your older mobile phones even as the communications backbone migrates from 3G to 5G. In energy, Tesla's use of modular battery solutions allowed them to build large utility-scale grid storage systems, and to deploy them with unprecedented speed. They were also able to continue to deploy new systems, even as the battery and inverter technology in the modules changed. Finally, the modular nature of the solutions also allowed deployment in smaller systems that are more distributed, providing similar capability at the grid edge. This move toward more modular and distributed solutions is a critical requirement for new solutions that stand a chance of being competitive in the twenty-first century!

We observe that because the economics of these twenty-first century solutions is so compelling, their adoption and growth seems to punch through regulatory and incumbent resistance. If the growth of these technologies is driven by individual companies, they will tend to want to create monopolies, where individual companies expand their market share using their own proprietary solutions. This by itself is capitalism and is not bad, but it can create friction and slow down the growth of a global market. Traditionally, industry cooperation and standards have been used to coordinate actions, but we have seen that there can be a 6–10-year lag that can make this very challenging, particularly as technology and knowledge multiply at a much faster rate than standards can manage. Other approaches may be required.

We observe that because the economics of these twenty-first century solutions is so compelling, their adoption and growth seems to punch through regulatory and incumbent resistance.

Fourthly, policy, standards, and societal ecosystems lag way behind innovation cycles, often resulting in chaos and causing transitions to slow down in achieving scale and impact. Of particular concern is that today we often regard market share and economic growth as the only metrics of value, and do not think holistically about the issues – triggering once again the concern of the "Tragedy of the Horizon" (to quote Mark Carney – "*the catastrophic impacts of climate change will be felt beyond the traditional horizons of most actors – imposing a cost on future generations that the current generation has no direct incentive to fix*") [13]. This almost makes these companies blind to critical and profound metrics such as life cycle costs, life cycle sustainability, carbon emissions, and waste management, concerns that can only be addressed through robust policy initiatives.

In our view, policy making and regulatory frameworks also need to innovate and be more forward-looking, iterative, and predictive to stay relevant. This should be the expected role of policy and government, who through the right proactive and predictive actions can help to accelerate market growth while achieving desirable holistic societal goals, such as: overall economic growth; reduction in carbon emissions; assurance that solutions are interoperable across manufacturers and technology generations and thus will not be stranded; addressing issues of equity and access; and ensuring that long-term costs such as end of life recycling and waste management are included in the solution costs and are not socialized. It should be noted that the above objectives are value-based and are technology-agnostic, allowing a variety of different technologies and approaches to compete. One way for governments to exert influence is through forward-leaning incentives (such as the German FIT program and US EV incentives) that become available at a point when the new solutions, all following a steep and sustained learning curve, are now nearing economic parity with existing solutions that do not meet societal goals. The incentives should only be available to those who are complying with the broader holistic societal objectives identified above.

Finally, we have also seen that rapid and uncoordinated growth in many adjacent and dissimilar segments can be a recipe for future societal chaos and potential disaster. Addressing this requires a holistic worldview and a forum to bring together the disparate sectors that normally operate independently. It also requires increased flexibility and resilience, and an ability to pivot and adapt to new realities as they emerge, and to not be hyper-focused on optimizing near-term profitability of one narrow segment. But, can this be done, and done fast?

Many of the energy transition-related technologies – PV, wind, storage, EVs – are all growing at unbelievable rates, and all need to connect to the grid (at least under the current paradigm). We see a strong move toward electrification of many things (possibly everything) because it offers the potential for lower cost, along with lower carbon emissions. For the first time, these lower cost exponential technologies have the potential to achieve our economic goals while positively impacting environmental sustainability. In other words, we can successfully address the highly divisive issue of climate change versus economic growth, not by being in conflict with each other, but by aligning the two goals of economics and climate! This is historic, because at no other time has it been possible to solve the challenge of economic prosperity while ensuring that our planet's living ecosystem can be sustained. As we will see below and in later sections, this transition and rapid coordinated change across the energy supply chain could prove to be very challenging but must be done right.

> *In other words, we can successfully address the highly divisive issue of climate change* versus *economic growth, not by being in conflict with each other, but by aligning the two goals of economics and climate!*

Even as energy transition technologies are growing at an exponential pace, unfortunately, the grid seems to be moving at its own pace, growing slowly, and reacting to changes only after they are seen to create major new challenges on their system. The grid is like the much-needed bridge for a successful transition. On one side of the bridge we have 1000s of gigawatts of next-generation clean energy sources waiting to come online, while on the other side, we have multitudes of clean energy users, ready to embrace it because it makes economic sense. But the bridge to connect the two is not set up to manage the traffic. Given the regulated structure of utilities, low competence in new technologies, low R&D expenditure, and slow processes for bringing new technologies onto its system, it is not clear that the electricity sector is positioned to act in a timely manner to meet the emerging requirements. The area of highest stress will be the grid-edge, where new generation and storage resources, potentially bidirectional loads such as EVs, and edge resiliency needs will change the fundamental paradigm of grid operation. This raises the question of whether the grid can cope with this change that is being thrust upon it.

Can we find ways to bridge the gaps, or will things fall apart? Will this lead to complete change in the way the electricity infrastructure operates, or will a hybrid structure emerge? Given that regulators tend to be unfamiliar with technology and tend to be political appointees, will they be able to guide the electricity industry well, helping meet societal needs and avoiding disruption. Or, will we go through a period of chaos, where cost of energy goes up even as reliability and resiliency suffer? This does not need to be the case, but absent proper steps and action, it very well could be.

If we look at how and why the energy transition has rapidly accelerated, it should be noted that reduction in carbon emissions and climate change have really been tangential to this discussion, and are positioned as very desirable, but only within the economic context. Given the Tragedy of the Horizon, we have, at a societal and global level, consistently shown that we are incapable of action, even to save ourselves, our children, or the planet, especially if such action does not provide near- to mid-term financial gain. The next section will look at how the next 20 years can potentially evolve. Given that we have been unable to predict how the last 20 years evolved, and that many more technologies are now following steep learning curves and are creating (or are poised to create) massive disruption in virtually every market sector, it is not clear why we will be able to predict any better now. However, we will try to provide a framework and foundation to assess what the outcomes could be, and what technologies, pathways, and investments we should consider. We will, in Chap. 6 and 7, also look at the possible role that new technologies and innovations can play in creating this future energy system and will also discuss where these new technologies and innovations could possibly come from.

References

1. https://www.intel.com/content/www/us/en/history/museum-gordon-moore-law.html. Also, Jafee, Amy Myers. (2021) Energy's digital future: Harnessing innovation for American resilience and national security. Columbia Press.
2. Fertina, N. (2022, August 26). *3D-printed solar cells are cheaper, easier to produce, and deployable at speed.* Interesting Engineering. https://interestingengineering.com/innovation/3d-printed-solar-cells-are-cheaper-easier-to-produce-and-deployable-at-speed
3. Azzopardi, O'Malley, Rajgor, and Richard, also https://www.telegraph.co.uk/news/2023/06/11/green-energy-disaster-uk-awful-warning-america/
4. DOE.org, https://www.energy.gov/articles/doe-releases-new-reports-highlighting-record-growth-declining-costs-wind-power (capacity factors of new land-based wind turbines).
5. *The Nobel Prize in Chemistry* 2019. (2022, June 7). NobelPrize.org. https://www.nobelprize.org/prizes/chemistry/2019/summary/
6. *BNEF, The Volkswagen Group,* for example, has promised 70 new EV models by 2028; 1 million vehicles by 2025; and 50 billion dollars invested in worldwide electrification. See Fischer, R. (2019, July 23). Also, NAR strategy presentation. National Academy of Sciences, Washington D.C., United States.
7. NASEM Report Finding 5.4 on the Future of Electric Power in the US, 2021.
8. Quote by Mike Howard, former CEO of EPRI (during the US NAE Committee on 'The Future of Electric Power in the United States'), 2021.
9. Metayer, M., Breyer, C., Fell, H. (2015). *Citation discussing industry inability to make accurate projections.*
10. Proceeding of the EPRI PES meeting, 2018.
11. Costello, K. (2016, May). *A primer on R&D in the energy utility sector.* (Report No. 16-05). National Regulatory Research Institute.
12. Ariza, M. A., Green, M., & Martin, A. (2022, July 27). Leaked: US power companies secretly spending millions to protect profits and fight clean energy. *The Guardian.* https://www.theguardian.com/environment/2022/jul/27/leaked-us-leaked-power-companies-spending-profits-stop-clean-energy
13. "Tragedy of Horizons", a YouTube video, Speech by Mark Carney at Llyod's of London, Tuesday Sep 2015 (BankofEngland.co.uk).

Further Reading

Brown, M. A., & Sovacool, B. K. (2011). *Climate change and global energy security: Technology and policy options.* MIT Press.

Clark, I. I., Woodrow, W., & Cooke, G. (2016). *Smart green cities, towards a carbon neutral world, a grover book.* Routledge Group.

For more on China's ascension in solar technology see pp. 145–6, Nussey, B. (2021). Freeing energy: How innovators are using local-scale solar and batteries to disrupt the global energy industry from the outside in. Mountain Ambler Publishing.

Jafee, A. M. (2021). *Energy's digital future: Harnessing innovation for American resilience and national security.* Columbia Press.

Nussey, B. (2021). *Freeing energy: How innovators are using local-scale solar and batteries to disrupt the global energy industry from the outside in.* Mountain Ambler Publishing.

5 The Next 20 Years—Utilities Get Ready for Decarbonization

Energy Transition Underway—Will It Be Orderly or Chaotic?

As we look ahead at the next two decades—from 2020 to 2040—the world, especially in matters relating to energy, seems to be positioned for massive change. By contrasting what happened over the past 6000 years, 100 years, and the last 20 years, we have shown a dramatically increased pace of change in nearly every sphere of life. In particular, we have seen rapid change and accelerated adoption across all business sectors for digitalization and "exponential" technologies. How the energy sector will evolve in the future is complex and difficult to predict and will be determined by the interplay of many forces including not only science, technology, and economics, but also policy and geopolitics. For instance, in developed countries such as the USA, the electric utilities could only achieve equity and universal access because they were highly regulated. Left to themselves, an unregulated utility sector would have left a large fraction of potential customers in the dark and unserved. Similarly, easily accessible oil and gas deposits were limited to only a few locations around the world, which made the petroleum sector highly sensitive to geopolitics. How will the next 20 years and beyond be different?

Almost by definition, the future grid and the future energy system will be here soon. Will this new infrastructure be what we want and need it to be? Can the irresistible force of new disruption overcome the immovable inertia posed by entrenched incumbents? New capabilities could help our society meet an aspirational objective of realizing an energy system that is sustainable, affordable, flexible, equitable, reliable, and resilient—or they could lead to chaos as two competing philosophies collide.

New capabilities could help our society meet an aspirational objective of realizing an energy system that is sustainable, affordable, flexible, equitable, reliable, and resilient—or they could lead to chaos as two competing philosophies collide.

D. Divan, S. Sharma, *ENERGY 2040*,
https://doi.org/10.1007/978-3-031-49417-8_5

How the future actually evolves will depend on whether the existing incumbent groups retain clout and slow down or prevent massive change from happening, or if new forces emerge that drive this massive disruption and change. Will this transition achieve climate goals and be economical, or will it be messy and costly and result in continued carbon emissions? Because of the complexity and scale of this transition, it is difficult to predict with certainty the path that will be followed in our journey to the future.

The recent NASEM report on The Future of the Electric Power System (Divan was a member of the committee that authored the report) acknowledges that over the last two decades, projections and predictions by leading organizations on the evolution of the energy system had been consistently and often spectacularly wrong. This has had big impact on governments, NGOs, as well as large commercial and industrial organizations—many of whom plan their investment and growth strategies over 10–20-year horizons, often based on these projections. The report concluded that a more reasonable approach may be to analyze multiple scenarios that could evolve through the interaction of the many distinct forces identified above, than to continue to rely on narrow predictions based on past experiences [1].

In this book we are also staying away from making predictions, even about the next 20 years, but we are very clear about our desires and expectations in terms of outcomes. As an aspirational goal, we would like energy to be abundant, clean, sustainable, reliable, resilient, and affordable—ensuring that its availability is not only economically viable but is also just and equitable.

> *We would like energy to be abundant, clean, sustainable, reliable, resilient, and affordable—ensuring that its availability is not only economically viable but is also just and equitable.*

With many simultaneously evolving technologies, differing regulatory and policy frameworks, and widely divergent national/regional energy strategies, many different paths will be taken. But the fast pace of change, and the accompanying uncertainty, also indicates that there will need to be pivots along the way, as better ways of doing something become proven and viable. This suggests that large complex infrastructure that takes decades and billions of dollars to build may have a difficult time competing with an alternative approach that is not rigid and brittle, but flexible, adaptable, and able to survive and bounce back from major misses in terms of where we are versus where we thought we would be. In the past, this would have been impossible and too expensive, but maybe there is an opportunity with twenty-first century technologies to realize such a system.

Given the complexity and interconnectedness of the issues and the rapid pace of change, can anything be done to manage the directionality of this change? Many of the technologies that are being touted as carbon-neutral solutions to our problems are far from the price and performance points needed to be directly competitive and meet today's market needs. It is also likely that additional major scientific breakthroughs may be needed to realize economically and technically viable solutions, for instance, in areas such as energy generation, storage, delivery, and utilization. To have impact over the next 20 years, the nascent technologies must first have critical translation issues resolved, then rapidly be commercialized, to finally scale by either augmenting or displacing existing (often incompatible) incumbent systems. It is also clear that properly timed forward-leaning incentives and policies can accelerate the adoption, but there are many questions surrounding this as well: are the incented solutions the right ones and what are the metrics; whose political influence caused the policies; and what are the adjacencies and externalities that are being ignored as we explore the impact of these technologies at scale? Also, as new scientific knowledge and technologies change at an unprecedented pace, can the solutions be flexible and adapt to new emerging realities on the ground?

This is the complex backdrop against which we need to analyze how the energy ecosystem will evolve over the next 20 years. It is not at all clear whether the energy ecosystem will be more electrified, more distributed and decentralized, more resilient, cleaner, with dramatically reduced emissions and abundant energy for all. Different ongoing research initiatives, entrepreneurial startups, and passionate discussions suggest that any or all of these outcomes are possible. If the right steps are not taken, we could have a chaotic system that is paused halfway through the energy transition (like deregulation of electricity markets in the USA), resulting in a system that is more expensive and less reliable than today, a system that needs coal and natural gas generation to back up poor reliability renewables and to charge electric transportation, where customers have solar panels on rooftops and EVs in the garage, but still have a dark house when grid power is interrupted by a forest fire. Major forces are at play, and the actual outcomes are anything but clear—but we have a chance to influence where we end up. Given that the goals we would like to move toward are broad and aspirational, and that human life as we know it may be at stake, we assume that unless their personal lives are negatively impacted in the near future, the vast majority of people would generally agree with these end goals. The challenge is knowing what such a future world might look like, and how we can get there without breaking the bank.

As discussed above, we have two major conflicting forces at work. The first is the group of incumbents: the electric utilities and the oil/gas companies (and petroleum supplying countries) who are in the process of being disrupted and will act to minimize the damage and impact to their organizations. Even if they say they want to transform themselves, it is very difficult to change the DNA and culture that makes the organization, and to walk away from billions of dollars in investments that have already been made.

The second group are the disruptors: the entrepreneurs, corporations, and investors who are bringing us PV solar, electric vehicles, batteries, hydrogen, and a host of other fast-moving twenty-first century technologies on a global basis. This is in turn driven by a global cadre of scientists, innovators, researchers, and technologists who generate and share knowledge with each other, adding to the global knowledge base. These disruptors are moving fast, and now that they are tasting success and feeling the heat of competition, they will want to accelerate even further. Funded by trillions of dollars in global capital (and thus clout), these forces are now also virtually unstoppable. While such Schumpeterian disruption has always accompanied progress and transformation, in this particular instance, if we follow the normal chaotic 50–100-year process for the transformation, we will not be able to act in a timely manner and prevent irreparable harm on the environmental front and to the quality of life for us humans as we move forward. *This timeline of change needs to be dramatically accelerated!*

Finally, because these are both regulated sectors with critical government involvement and oversight, regulators and lawmakers will also have a big influence on the goals that are set, mechanisms to seed-fund, incentives to accelerate early deployment, and regulations that can enable scaling in a timely manner—thus influencing outcomes at a societal level.

There is thus an urgent need to at least have an informed dialogue about what our options are, and the consequences that will possibly result from the various choices that we will make. There is an opportunity to identify those technologies and solutions that are aspirational and aligned with our end goals, in terms of delivering value, and have the potential to be economically advantageous, while creating policy that can accelerate the adoption and scaling of these solutions. To achieve this, we have to better understand the many forces that are driving the process today, as well as the challenges, roadblocks, and opportunities. We need to look at all of this in a nuanced manner so we can understand the interactions between the various sectors and initiatives and see where some of the nonobvious challenges lie.

Will a Decarbonized Energy System Be More Electrified?

There are many who continue to question that renewables could ever scale to replace fossil fuels as a primary source of energy. Renewables in 2019 represented only 6.4% of the total energy generated in the USA—a 16X growth seems impossible. On the other hand, as Fig. 1.1 shows, of the 93.6 Quads of fossil or nuclear energy used (both using thermal energy conversion principles), only 26.1 Quads or 27.8% of the energy was converted into actual end use. Fully 72% of the energy generated from thermal processes was rejected as heat. Put another way, for every dollar we spent on extracting, processing, and delivering energy, we threw 72 cents away! With renewable generation and electrical processes, energy conversion efficiencies are high, often between 90% and 99%, and little of the generated energy is wasted. As we replace wasteful processes that were developed hundreds of years back, with more efficient processes that run on electricity, we suspect there will be a move toward increased electrification—first for transportation, then for basic industrial processes, and finally at a societal level.

As a result, we feel that the fossil fuel sector will taper down and see a diminishing role in our future energy mix, while electrification and the associated power grid and the pathways to deliver this energy will form a key part of the infrastructure that ties many of these elements together, and therefore is likely to play an even more important role than it does today. PV and wind generation, energy storage, and electrified transportation are all fast-moving, relatively unregulated and highly competitive industry sectors, all seeing falling prices and explosive growth across the globe.

As EV populations increase at a 60% year-over-year rate, they will expect to receive as much energy as they need, wherever and whenever they connect to the grid! Similarly, PV solar and wind generation, which are also growing at explosive rates and can provide inexpensive energy but only when it is available, also expect to connect to the grid and, like magic, have their generated energy delivered to their customers in a manner that maximizes their own revenues. Also, we cannot forget the millions of existing residential, commercial, and industrial electricity customers, who all expect to continue to receive electricity, with even lower rates than they currently pay (because solar is now cheaper, right?), and with at least comparable levels of reliability and even higher resiliency than they enjoy today. Finally, we expect to see customers with new loads emerging, driven by a wave of increasing electrification, who will also expect to have access to electricity, wherever and whenever they want it. But can the grid evolve at the pace needed to be what each and every one of us wants it to be?

While prices of utility scale PV and wind energy are now below conventional dispatchable resources on an energy delivered ($/kWh) basis, that may be too narrow a metric as we look forward. Because fossil fuels are burnt at the time of use and have an energy output that is directly related to volume, and because the extraction, processing, delivery, and storage infrastructure represent fixed costs that can be amortized over a long time, it is understandable that costs at point of use largely relate to the cost of delivering the fuel to the end use location (e.g., to a generation plant or an automobile). As such, a cost of energy that is based on energy content of the fuel is appropriate.

When we look at PV solar, all the cost is in setting up and maintaining the generation plant, with virtually no added cost for actually generating electrical energy—the marginal cost of the next kWHr of solar energy is zero when the sun is shining. With declining PV prices, a long plant life of 20–25 years, and minimal operating costs, it is not surprising that the cost of energy delivered from a PV solar plant is very low and decreasing. If we were prepared to take this energy, whenever and wherever it was available, we would undoubtedly have the lowest cost of energy.

However, that is not the case. We also want energy when the sun is not shining and the wind is not blowing, sometimes for weeks at a time. What's more, we do not want to really cut back on our energy consumption at times when the sun is not shining for extended periods, begging the question of where that energy will come from. By way of example, in India when the monsoons arrive in June/July, there can be months when the sun is blocked by clouds over large parts of the country. How will a PV dominant system operate at such a time? Even long duration energy storage will not cover this gap. We have to maintain (or build) an alternate dispatchable base generation resource from which the load can be supported for weeks or even months. Further, this dispatchable generation base will only be used on the grid when the low-cost renewable resource is not available, and as such will be very expensive on a $/kWh basis if the plant's operating costs are to be fully covered by the grid services it provides. This also implies that at such times of energy shortage, the actual cost of delivering energy in real time could be very high. This also suggests the need for dynamic pricing and load prioritization right down to the grid edge, as a possible mechanism to balance the grid in real time.

Today, we are not prepared for this. Most consumers and small to medium commercial and industrial customers in the USA see flat electricity pricing, with some utilities using time of day pricing as well as a very primitive average-load balancing mechanism. In those cases, where unrestricted market operation has been allowed (e.g., Texas), electricity prices that some customers paid during the 2021 polar vortex approached $9500/MWHr, as opposed to a normal price of $120/MWHr. Even if the customers were aware of the increased real-time price (which many claimed they were not), they had no other options because there was no mechanism for them to disconnect from the grid and to use lower cost stored energy for their higher priority loads, such as lights! The only option was power at $9500/MWHr or no power at all!

While we customarily accept the surge in pricing for our "Uber" rides at peak hours, see dynamic pricing for airline tickets during holidays, or high prices of gasoline when Russia declares war on Ukraine, we are not accustomed to wide swings in electricity prices. There is a real need for dynamically managing demand on the grid and having it follow available generation capacity, using pricing and microgrids that the utility can disconnect as needed to balance the system. But it is equally clear that the need for dispatchable base generation that can run for extended periods of time will likely not go away.

As we have observed before, the variability in energy price will also adversely impact low-income communities disproportionately. We have already seen extreme examples of this in the case of HILF events, when poor communities had to go without power for many weeks. As we will discuss more in later sections, *any solution that is proposed should include considerations for equity and should ensure that a basic minimum level of service is available for all citizens, even under the most adverse conditions.*

So, what do we really want from our energy infrastructure—and can this be realized over the next 20 years, even as we reduce our carbon emissions and try to meet our climate change goals? Do we have all the technologies and solutions we need? Some aspirational needs and some big gaps can be identified. Electric vehicles seem to offer a good option for meeting transportation needs, but only if we can ease concerns about range anxiety and solve the ubiquitous rapid charging problem. As we have seen, this can have a big impact on the grid. We would also like to steer ourselves away from carbon emitting fossil fuel-based generation and shift to renewable energy. Storage can take care of a significant part of the dispatchability problem but is unlikely to meet the challenge of extended run time for weeks to months. This requires on-demand generation that can come from hydropower, nuclear energy, or from hydrogen that is made from renewable energy or from fossil fuels with permanent and safe sequestration of carbon and waste streams.

We would like to shift our entire energy usage to these lower cost cleaner energy resources, even as more of the industrial sector shifts from fossil fuels to electricity.

All these sources and loads will need to connect to the grid and will need to be balanced and managed. At the same time, the question of resiliency and equity will need to be addressed—again a grid issue. Finally, we must address the question of who will pay for this new infrastructure that needs to be built. The grid will clearly play a vital role in the future energy system—but only if we transform it to fit our future needs.

The grid will clearly play a vital role in the future energy system—but only if we transform it to fit our future needs.

As discussed in earlier sections, electric utilities are highly regulated, have a very low risk appetite and are unable to rapidly respond to change. One can see that this situation is possibly manageable at low DER penetration levels (because today's grid is significantly overbuilt), but it is very likely that the incompatible rates of change will eventually create unacceptable levels of stress at the grid edge. How this stress will manifest itself is not well understood and is perhaps the trillion-dollar question and could have a very big impact on how the scenario unfolds and will thus be a significant part of the discussion in this section.

Electric Utilities Get Ready for Decarbonization

That this transition could happen is not news and has been discussed for many years by the electricity industry, the DOE, national labs, and regulatory organizations such as FERC, NERC, and NARUC, as a possible, but at best, a slowly unfolding scenario.

At the EPRI Summer Meeting in 2021, leading electric utilities finally acknowledged that change was coming like a freight train and could not be stopped, and in response, committed to a carbon reduction goal of 50% by 2030 and 100% by 2050. As a result, many utilities are no longer publicly and actively resisting change (at least at the leadership level) and are beginning to say that they need to act, and to anticipate possible major disruptions to their systems, whether driven by changing customer demands, regulatory fiat, or new technology. This transition will not be easy!

We must recognize that utilities, working with their regulators, are acting as good stewards to preserve their organizations and to execute on their current charge—providing reliable, affordable, and equitable energy for all. The new requirement of meeting 100% carbon reduction by 2050 can be perceived, at least in the early years, to actually reduce system reliability, affordability, and equity—in direct contradiction to what their traditional goals are! This is the reason why utilities have pushed back on the renewable-energy revolution for so long. But now that utilities are committed to the energy transition—how are they viewing the pathway to this transformation? It is interesting to see it directly from their perspective.

The EPRI 2021 reports [2] released at the above meeting represent a consolidated and considered perspective of how electric utilities are viewing the energy transition and provide perhaps a good starting point to understand what they think they need to do and where the gaps are, if any. The reports describe the following key elements (paraphrased for succinctness) that need to be addressed by utilities as they look at the upcoming energy transition, including:

- *Resource Adequacy*: Availability of sufficient generation resources and ability to deliver energy where needed to meet demand at all hours and locations, under all conditions, including events such as equipment failure.
- *Transmission and Distribution Infrastructure*: Sufficiency of transmission system capacity to enable interregional economic energy flows and intraregional delivery from generation hubs to load centers. Sufficient distribution system infrastructure to enable distributed generation, serve increased demand from electrification, and allow flexible behind-the-meter sources and loads to provide grid services—all while maintaining distribution reliability standards.
- *Operational Reliability—Grid Stability, Balancing, and Flexibility:* Ability of the grid to maintain desirable and stable system performance under a variety of operating conditions and credible disturbances and faults, to prevent cascading outages, and to ensure reasonable restoration for disturbances that are beyond planning criteria. Balancing and flexibility require that the grid is able to instantaneously balance supply and demand under all operating conditions, including normal, transient, and abnormal conditions.

The above elements only include operational issues and not the economic, regulatory, and policy-related factors that are perhaps equally, if not more, important, and which are also acknowledged in the report as being critical. On the surface, this approach seems to reflect a thoughtful and balanced pathway for what the industry needs to do to sustain its relevance. The devil, of course, is in the details. It will be illuminating to understand how each of the above factors can be unpacked to show specific actions that need to be taken, expected outcomes, and what the challenges are likely to be.

Resource Adequacy

As we have seen in earlier sections, today's grid operates in a centralized manner based on continuous bulk system monitoring and state estimation with generation dispatched to minimize cost while keeping the system stable and all equipment within operating limits. This is still the model that most utilities are pursuing. Now, however, hundreds of gigawatts of distributed generation are being added to the mix—often in locations where the solar or wind resource is strong and economical but is also likely to be far from locations where the energy is needed. Ignoring for now the question of new transmission to connect these large utility-scale generation plants to load centers, we need to worry about how well resource availability matches demand.

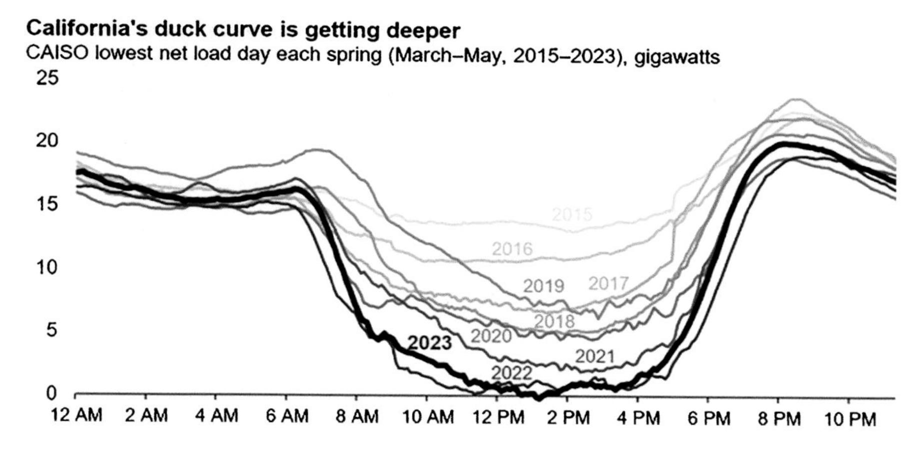

Fig. 5.1 California "Duck Curve" showing impact of increasing PV solar generation over time in a day, including periods of overgeneration, and steep ramp rates in the evenings (https://en.wikipedia.org/wiki/Duck_curve)

The mismatch is well captured by the California "Duck Curve" [3] (Fig. 5.1), which shows that as solar generation increases, we see significant gaps emerging, including generation/demand mismatch and "ramp-rate" requirement of 10 GW/hour for the generation resources that need to step in as solar energy diminishes at the end of the day. On top of this, of course, is the question of what happens on cloudy and rainy days (and weeks). If we add wind energy to supplement these gaps, we can see an improved balance, on an average, between resource availability and demand, leading to several academic papers that posit that the USA (and the world) could get to a 100% renewables-based system with no issues [3]. As we have seen, meeting the demands on an average day is not enough. The load also needs to be supplied on those extreme days and weeks when the sun does not shine, and the wind does not blow.

For many utilities, the 2030 target of 50% reduction in carbon emissions can be managed simply by turning off a few coal plants (that were already scheduled for closure), especially if the selected reference year for measuring reductions is in the past when emissions were predominantly from coal plants and were very high. The USA has already seen a significant reduction in carbon emissions as natural gas from fracking has become competitive, which has more to do with sensible economics than reducing emissions (it is perfectly fine for utilities to count this as a win—after all the objective is achieved!). Gas turbines were thus rapidly deployed to primarily meet base load, but to also provide the peaking capacity needed to fill in the gaps in renewable generation.

However, as utility scale solar and wind today already cost less than coal and gas in most places, gas plants can no longer be competitively operated at capacity factors that ensure their economic viability under the current rules and pricing. In any case, they do generate carbon dioxide, albeit less than coal plants do. As such, they are not the long-term (2050) solution to the carbon neutral problem but may be an excellent short-term solution. Further, with appropriate rules and incentives, new emerging technologies such as Direct Air Capture coupled with permanent sequestration of captured carbon, can provide the mechanisms for ensuring a soft-landing to a zero-carbon system, but are still too expensive today.

Other mature non-carbon resources that are already available at scale include nuclear and hydropower. For reasons discussed in Chap. 4, traditional nuclear is in decline, with an accelerating pace of continuing shutdowns of aging plants that are well past their design life. Further, unresolved issues such as plant and environmental safety under extreme events, and viable solutions for long-term disposal of nuclear waste are leading to continued community pushback. Finally, new plants like Vogtle are seeing extreme delays and cost overruns, thus calling into question the economic viability of today's nuclear plants and technology as a long-term future energy resource. However, there is talk about new nuclear technologies, including small modular reactors, which could change the equation—but they are not ready for market yet, and long-term issues of spent-fuel management have not been solved.

Looking at hydropower, we see that it is clean and carbon neutral but is also very site dependent and often faces challenges in terms of pressure to prioritize water for human consumption and irrigation, as well as environmental factors and community pushback. Some of the older hydro projects are approaching end-of-life, and it is unlikely that many new hydropower projects will see fruition in the USA. For an increasing number of utilities that are either facing RPS mandates or are on the hook to meet the lower emission standards, meeting the demand without dispatchable generation that can operate at very high capacity factors and at will, poses a very big challenge.

Recent events in Puerto Rico, California, and Texas point to additional challenges with resource adequacy, especially when dealing with extreme weather, fire, or other events. With increased penetration of energy infrastructure, urbanization, and population growth over the last century, our dependence on the electricity infrastructure has only increased. As a result, power outages that leave communities in the dark for weeks, even months, have a disastrous impact on daily life. Unfortunately, these events are clearly observed to be increasingly common. Additionally, the consensus among scientists is that these energy disruptions will be further exacerbated by shifting weather patterns, caused by changes in atmospheric and oceanic streams, and their related impact on overall climate.

This is the case when traditional dispatchable generation is available on the grid. Now consider a scenario where the predominant source of energy is wind and solar and we have to additionally factor in extended periods of cloudy/rainy weather that result in the loss of the entire PV generation fleet for a state like California. Or consider wind energy. Yes, wind has a higher capacity factor than PV, but it can also shut down due to a lull in the wind, or due to strong winds that trigger protective mechanisms (as happened in Texas in 2018, almost resulting in a cascading failure). These challenges once again raise many questions. How does a utility plan its operations with such variability, especially when they have a renewable energy dominated grid? How do they guarantee they will meet demand? Are customers ready to accept lower reliability, especially for certain load segments, simply to be green? Sure, we can overbuild the generation and storage assets so that every contingency imaginable is covered. But what would the cost of such an overbuild be? Would energy still be economical? Who would pay for the overbuild?

Energy storage seems to be part of the answer. Today, so as to improve grid integration of renewable energy, there is an increasing interest (with many early installations) of hybrid plants, which include generation (such as PV) with energy storage. Batteries seem to be the most economical storage resource today, with typical systems offering 2.5–4 h (and even more) of energy storage. This allows a modicum of dispatchability and grid integration and allows these resources to participate in the market but does not cover the cases where days or weeks of energy storage are needed. Even batteries for grid storage, although being deployed at more than 100 GW/year, are still relatively new without a clear validation of life in the field under real use conditions and are still plagued by safety issues such as fires and disposal at end of life. Even assuming batteries will prove themselves, we still need a solution that provides days to weeks of backup power and is clean—which is simply not available, especially at the scale and cost point that is needed!

The second half of the adequacy equation is load and demand. If we look in the rearview mirror, the situation looks reasonable, because load is increasing very slowly. On the other hand, if we look ahead, we see that to move away from carbon emitting fuels will also require a move toward increased electrification. Many cities are signing ordinances banning gas for cooking in new homes that are being built, requiring the use of electricity. Similarly, many industrial processes are being electrified. Other transportation sectors, from trucking and rail to ships and aviation, are looking at electrification as a possible path to carbon neutral operation. Hydrogen is being considered for long-term energy storage and as a fuel for generation and heavy transportation, with "green hydrogen" requiring electricity for its production. As can be seen, many of these potential solutions also intersect with the grid and will increase demand for electricity even more.

It should be remembered that in much of the USA, generation is not controlled by the utilities, but by an Independent Power Producer (IPP), who can take on technical and commercial risk and can choose to move as fast as they want. This is happening, for example, with the surge in PV solar, wind, and storage projects in recent years. However, to succeed, they have to be able to interconnect with the grid, with a service agreement that gives them a reasonable chance to be profitable. The interconnection request is in the hands of the utilities and grid operators, who may not have an equally pressing mandate to move fast. In fact, from their perspective, rapid and uncontrolled growth can impair system reliability and would not be desirable. As a result, interconnection queues have grown dramatically to over 2000 GW, with estimated times of over 4+ years in many places. Some of the largest grid operators such as PJM have put a 2-year hiatus on any new PV interconnection applications. The reasons for this will become clear as we look at power delivery, which includes transmission and distribution (Fig. 5.2).

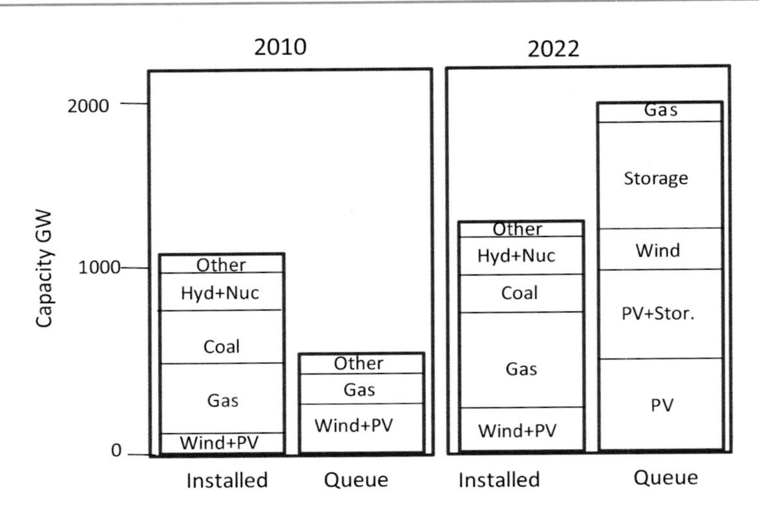

Fig. 5.2 Transmission interconnection queue at the end of 2010 and 2022 (LBNL) [4]

Power Delivery: Transmission and Distribution

If we stay with the current paradigm of "lower cost" utility-scale, centralized bulk generation as we move toward a greater level of electrification and simultaneous decarbonization, we will invariably need more transmission lines to enable interregional economic energy flows and intraregional delivery from generation hubs to load centers. This includes AC transmission at 230 kV, 345 kV, 500 kV, and even 765 kV to transfer power between states and regions, as well as HVDC lines to transfer large blocks of power over long distances between regions or from large generation regions (hydropower in northern Canada or large PV solar in Arizona) to distant load centers.

The need for more transmission has been there for decades, but growth has been slow because of regulatory roadblocks and NIMBY ("not in my backyard") concerns. On an existing base of around 240,000 miles of high-voltage transmission lines in the USA, we have been adding around 1000 miles per year (0.25% growth rate) or so for the last 20 years, when the projected need is for a 57% increase in transmission capacity by 2035, a goal that appears to be very challenging [4, 5]. Given the community pushback, it is not surprising that it has often taken decades to get new lines built—yet it is a critical part of what is needed if the current paradigm is to continue! FERC and NARUC are realizing the urgency and are ramping up the pressure, but execution is often hampered by politics. New initiatives are being introduced to promote cooperation between regulators from various states and to use eminent domain to drive access to transmission corridors. But this is likely to be a slow process and faces great challenges.

As we look out over the next two decades, we see that the rapid pace for building new renewable energy and storage plants and shutting down fossil generation, especially in the USA, is out of sync with the glacial pace of building the new transmission lines needed to meet energy demand—both for multistate projects as well as for smaller intraregional transmission upgrades. As discussed above, although PV, wind and solar farms can come online very quickly, the timeline for building new transmission, a critical part of the needed infrastructure, is severely mismatched. Although the reasons for these delays and challenges are innumerable, ranging from environmental reviews, permitting, cost allocations, regulatory issues, and financing—*perhaps the most worrisome is that changes in the renewables dominated marketplace have raised the financial risk profile and made investment recovery more challenging for transmission investors and operators*. These delays further impact an already challenging concern of economic viability. Finally, the poor capacity factors of the new generation resources lead to transmission line utilization that will be even worse than for existing transmission lines, already very poor because of redundancy requirements, making economic viability even more challenging [5].

Finally, we suspect that the economic realities of traditional technologies versus the new exponential technologies will also prove to be very challenging. For instance, a new 345 kV transmission line that can move say 400 MW of power, costs around $2–2.5 M per mile (and increasing every year), with additional costs for land access and substations, but can only move power over 50–100 miles and can take 10–20 years to permit and build. Longer line lengths of up to 200 miles are possible for AC lines but would require higher voltage lines or additional costs in terms of dynamic VAR compensation, ratcheting up the overall cost of a 200-mile line to >$500 M. These new lines would not have any dynamic control capability to handle the variability of renewable resources. Today, such capability would require HVDC, especially for delivering power over long distances (50–1500 miles) along overhead lines, or even shorter distances with cables that are buried in soil or underwater (10–100 miles for offshore wind projects).

A Voltage Source Converter (VSC) based +/−250 kV HVDC line of 200-mile length that can move 500 MW of power is estimated to cost around $550 M, including the converter station, again with an expected timeline to operation of 10+ years, given the issues of transmission line permitting, siting, and build. Because the cost of utility PV solar is already <$1/watt and continues to decline with a learning rate of 24%, it may make more sense for the PV farm to be built closer to point of end use, even if the cost of the farm is higher, say by 20–30% or so. Total new PV farm assets of 400 MW could be geographically distributed as needed, would cost less than $400 M (and decreasing every year), and could be sited closer to loads such that new transmission, with all the permitting and other issues (and an additional cost of $500 M), may simply not be needed. As we have seen, it is expected that over the next few years, the cost of PV solar including 3–4 h of energy storage will approach $1/watt, allowing significantly improved integration with grid operations than is possible today with a pure PV farm. There is a different set of issues for densely built urban areas, which will be discussed later.

The above discussion highlights the challenges that a traditional expansion of the transmission network is likely to face. Actual targeted expenditures on new transmission lines are significantly short of what is needed—for instance, the 2021 $1 T infrastructure bill only contained $2.5B specifically for transmission, versus the estimated $360B needed to get to a carbon neutral grid. Even with the best of intentions, it appears that new transmission will take time to build and may not be available as the new renewable energy resources are deployed at scale. *This further exacerbates the tension between traditional approaches that move slowly and deliberately and where costs are increasing, and the new technologies that are rolling out at supersonic speed with steadily declining costs!*

Based on the above discussion, it appears that an approach that should perhaps be considered is to use existing transmission assets, most of which are already depreciated, and to use dynamic control or capacity expansion technologies that are already available, and which can allow us to improve utilization of the existing transmission system. In addition, we may want to reduce our dependance on building a lot of new transmission lines, focusing resources instead on the few critical new lines where capacity improvements provide the most impact. Such an approach would provide very specific guidance on where new renewable generation resources should be located, would help prioritize investments that improve utilization of existing transmission assets over building of new transmission lines, and would incentivize location of new generation on the distribution grid closer to points of use.

Under the existing grid paradigm, the distribution network connects the bulk power system with all end users. Power flow is unidirectional, with a radial network going from a substation along a distribution feeder to the end user. Even as US utilities focused on bulk power system reliability as a key metric, achieving 4- and 5-nines in uptime, their actual customers at the grid edge experienced reliability levels that were much worse. Over the last 10–15 years, utilities have begun to slowly upgrade their distribution systems, adding Distribution Management Systems (DMS), smart meters, sensing and measurement devices, and sectionalizing devices such as “Intellirupters” to improve reliability—but at best, this remains a work in progress! Most utilities still tend to have very poor real-time visibility over their distribution systems, and even poorer control capability. One of the biggest challenges for them is the explosive growth of DERs, and the new loads arising from electrification of traditionally nonelectric loads.

The rapid growth of PV solar energy on the distribution network, even at low penetration levels, has caused great concern for most utilities. Stated concerns include voltage volatility, protection relay settings due to reverse power flows and fault current contributions from renewable resources, anti-islanding protection that should inhibit grid-connected inverter operation when the grid is down, and a billing structure that penalizes poor people who cannot afford PV solar systems. Strict guidelines for grid interconnection of PV systems have been enforced to slow down the growth of renewable generation on the distribution system. At the same time, EV penetration is rapidly increasing, and along with that the patterns of loads on the distribution network are changing. However, the fact that electrified transportation is growing (from 1 M vehicles in 2018 to a projected 125 M by around 2040), raises concerns about how the grid will cope with the new loads as they come online.

The EPRI report succinctly, and at a very high level, presents this as a need for new distribution infrastructure over the next two decades to ensure that distributed generation is enabled, increased demand from electrification is served, and flexible behind-the-meter sources and loads can provide grid services—all while maintaining distribution reliability standards. The report talks of a transition from a *DER agnostic grid to a DER dependent grid*—but leaves unsaid a lot of the issues that such a transition represents. For example, customer sited solar grew from 1 GW in 2010 to 32 GW in 2020 and is projected to grow to 320 GW by 2050. In another study, DOE estimates that to get to 0% carbon by 2050, we will need to have a total of 1000 GW of solar by 2035 and 1600 GW by 2050 (which, based on today's ratios, implies 1300 GW of utility solar, all needing transmission!).

We have also seen the impact that fast charging across the USA can have on the required peak generation capacity. In view of the previous discussion on challenges with deploying transmission, and with all the challenges of permitting, NIMBY and the financing of grid extensions, it is difficult to see how the slow and highly regulated utility planning and build process could allow such rapid growth in the transmission and distribution infrastructure to occur. These are massive changes at an unprecedented scale and will be very challenging for an industry that has moved slowly and deliberately for the last 75 years!

> *These are massive changes, at an unprecedented scale and will be very challenging for an industry that has moved slowly and deliberately for the last 75 years!*

As discussed in Chap. 3, we see that another major related issue is how to coordinate millions of non-utility owned PV inverter assets (Inverter Based Resources or IBR) so that the grid operates as desired. There are several layers to this problem. Unlike utility owned generation, this new resource is privately owned, with high diversity of owners, manufacturers, technologies, technology generations, communications capabilities, and intelligence at the edge. Standards to which these PV inverters are designed, and presumably comply, have been shifting dramatically as utilities have gone through changes in their perspective of what they want these inverters to do (and the change is still continuing!). The vast majority of inverters deployed today simply follow the grid voltage and pump power as it is available, relying on net-metering to reduce their cost of energy, or to be paid by the grid operator.

If we now consider interactions between the grid and the IBR devices, at low IBR penetration levels, it does not really matter, as the grid is strong enough to absorb any irregularities that individual inverters may impress on the grid. However, as has been discussed earlier, as IBR penetration increases to represent a higher percentage of system capacity, the inverters have to work together to actually form the grid, not just to follow it. This is a process that requires millisecond level coordination (something synchronous generators intrinsically do, but inverters do not).

This may be feasible when all the inverters are from the same manufacturer, are collocated together, and are custom designed and installed to be able to share power and operate in tight synchronism—which leads to reduced flexibility and innovation.

When you have a diversity of inverters with different manufacturers and owners that actually need to be coordinated, this seems to be an impossible task! When you factor in the ownership, communication, and cybersecurity challenges, we see that this can become an intractable problem. Grid forming inverters are still evolving and face many difficult challenges and are possibly 10+ years from deployment at scale, given today's lagging standards process. Organizations such as the DOE funded UNIFI consortium are working on addressing the issue of grid interconnected IBRs [6]. However, during this interim period, while a suitable standard is being developed, millions of inverters that do not comply with a possible future grid-forming inverter standard will already have been deployed!

To ensure that these IBRs are coordinated at the market layer with grid needs, utilities are turning to providers of Distributed Energy Resource Management Systems (DERMS), who are aggregating these IBRs in a utility's service territory, with the goal of coordinating the collective response of these resources as a "virtual power plant" and integrating them with the market. This works reasonably well for larger commercial and industrial-scale resources, because the cost of integrating and operating the communications and cybersecurity overhead is a small part of the overall system cost. At the individual residential and small PV system level, integration with a DERMS provider can prove very unwieldy and expensive, especially given the lack of standardization on both hardware and software. The smaller residential systems can also provide a cyberattack pathway into the grid system, especially if the residential PV system is connected to the home WiFi network. It has also been proposed that the smart meter could provide the gateway device that can control and manage the residential PV system. However, the complexity of integration and cyber vulnerability of such an approach is possibly worse than with a DERMS provider.

The new resource that is now becoming available, even on the distribution system, is energy storage. Storage can be owned by the utility and used to support grid functionality. Or, it can be owned by the end user, and be used to support their internal needs—such as peak demand reduction, energy arbitrage, or back-up power. Storage has a key role to play in the future grid and will be discussed in more detail later. Even though the EPRI report mentions grid-support from customer owned resources, regulations in most states prohibit "behind-the-meter" (BTM) resources from participating in grid control functions and from benefiting financially from providing such support. Further, even if we could get the regulatory issues resolved, we do not currently have standardized technical solutions that are widely agreed upon to address these issues in a truly scalable manner. This issue will also be discussed later.

Operational Reliability—Grid Stability, Balancing, and Flexibility

The previous issues involve generation and power delivery and are largely in the slow "transactive" domain—having to do with slow phenomenon and processes, such as optimizing system cost and performance, achieving grid market objectives, ensuring that the right investments are made, and that system planning ensures high reliability and lower cost. The grid operator computes a "security constrained economic dispatch" (SCED) for each generator based on the system state and objectives and sends the updates out every 1–15 min. At the transactive and business level, this is all we care about.

However, there is more to grid operational reliability than system awareness and economic dispatch. Maintaining an operational grid, at least until now, has required keeping thousands of generators spinning in tight coordination with millisecond precision, and having the system operating and delivering power in a reliable and safe manner to millions of customers. The generators need to, both dynamically and in the steady state, control their power levels to balance supply and demand under all operating conditions. The grid also needs to remain stable under all normal and abnormal operating conditions, and allow quick restoration when conditions are beyond the design parameters.

Under the current paradigm, utilities are addressing these issues by improving awareness of grid operating conditions and grid reliability using selective deployment of advanced instrumentation and protection devices, such as Phasor Measurement Units (PMU), intelligent relays, transformer monitors, advanced metering infrastructure (AMI), breakers, sectionalizers, etc., to ensure that the grid operates within nominal parameters, and does so in a stable manner. And grid reliability has gradually improved, and things seemed to be on the right track—until DERs started showing up on the grid at scale!

It is important to understand how physics plays a role in grid operations and stability. The above discussion suggests that the grid operator is in charge of everything that happens on the grid. But the grid operator only issues an update every 1–15 min. In between these updates, the grid still needs to keep operating, even as the loads change, and contingencies, faults, and failures occur in an unanticipated and uncoordinated manner. The generators are operating at 60 Hertz, with one electrical cycle completed every 16.6 milliseconds—which implies that with a 15-min update rate, 54,000 cycles have to be automatically managed by the grid with no active control from the grid operator. *To put that in perspective, if we received guidance from, say a financial advisor, on what daily actions we should take based on market conditions—we would be waiting for 148 years before the next input was received!* Clearly, today's grid automatically keeps on operating, purely based on the laws of physics and the control rules that are in place. It is a testament to our engineering ancestors that they could design such a system, well before we had microprocessors and computers. But are the same operating principles adequate for the new future grid?

For most of us (including many attached to the utility industry), our intuition, which is based on a largely physical and mechanical world that we perceive and personally interact with, completely fails us when we look at the electricity grid. Regulators, many of whom are lawyers and economists and have possibly forgotten any physics they may have learned in high school, often struggle with ideas such as the need to achieve instantaneous balance between generation and demand, the concept of reactive power, or why energy put into the grid at Point A at a specific time cannot be consumed at Point B at the same or different time, and complain about how "physics" interferes with their economic and political objectives! For most people, the process by which the grid operates is invisible and is taken for granted. Because the system has worked reasonably well for 100 years, they assume that it will simply continue to operate, even as the sources, loads, and the grid itself change dramatically. As a result, a lot of the discussion that occurs tends to focus on slower functions such as market, regulatory, and now cybersecurity issues, or the need to comply with irrelevant, or even outdated standards.

The underlying assumption often is that the intelligence largely lies in the sensors and the "slow" "higher-level" market layers, and that the "lower-level" execution layers are purely electromechanical with time-invariant controllers—which was largely true for a grid based on electromechanical synchronous generators but is definitely not true for inverters. This also suggests that in today's conventional grid, only the controls at the grid operator (separated from the grid by 5–15 min) are smart—the grid itself is largely electromechanical and passive with no intelligence. Inverters, on the other hand, can react to inputs and disturbances in microseconds and incorporate significant intelligence, as well as computational and analytical capability at the grid edge, which allows them to perform a host of functions, and to change their behavior dynamically as circumstances change, just as we humans often do!

This is both good and bad: good, because inverters can act fast to protect themselves and react to unwanted disturbances; bad, because when many inverters act fast, they can interact with each other and cause grid instability issues. This suggests that future inverter-based resources (IBR) would also need to autonomously meet "operational requirements" for the future grid, essentially ensuring that the system is stable, balancing generation and load everywhere and at all times, and having the flexibility to be able to respond to unexpected conditions on the system, including transients and faults. Figure 5.3 shows a representation of transitioning from today's

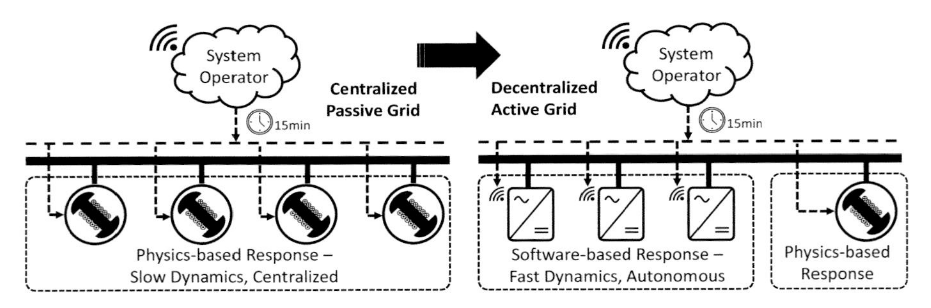

Fig. 5.3 From passive grid to active grid with intelligent agents at the grid edge for dynamic control

Smart Grid, which is essentially passive and electromechanical with a slow overlay of optimization and intelligence, to an Active Grid, where millions of intelligent agents (inverters) at the grid edge act *dynamically in a collaborative manner* to realize a stable grid. While, this is a compelling representation, it remains an aspirational goal, one which we do not have the means to get to, just yet.

The future IBR-rich grid will be very complex, with millions of devices interconnected by electrical power lines, with communication links that have latencies and cyber-vulnerabilities, and many layers of regulatory compliance and economic objectives that need to be simultaneously complied with. Isn't this the norm today? We have large communications and computing networks that span the globe—so why is the grid different? The major challenge and the potential for wide area cascading failure arises from the real-time physics of coupling large generators, millions of inverters, and other system elements. The grid represents a "real-time must-run" (RTMR) system, where every generator and grid element must automatically keep functioning, delivering gigawatts of power to millions of locations with millisecond precision for 54,000 cycles (15 min) before the next update is received, and must operate even if it is not!

The grid represents a "real-time must-run" (RTMR) system, where every generator and grid element must automatically keep functioning, delivering gigawatts of power to millions of locations with millisecond precision for 54,000 cycles (15 minutes) before the next update is received, and must keep operating even if it is not.

Unlike a data or telecommunication system, this system cannot be paused for a few seconds, or even milliseconds, while a cyber issue is resolved, or signals are rerouted because of a failed or overloaded router. Such a delay, which we are very familiar with in our televisions and cellphones, would be fatal on the grid and could result in the wide area cascading failures and blackouts that we would like to avoid. We have operated such a RTMR grid for 100+ years with high reliability, mainly because the large synchronous generators have substantial rotational inertia and damping that helps keep the system in synchronism, under all but the most extreme fault conditions. This changes substantially in an IBR-rich grid.

As we look at the possibility of taking a centralized control paradigm that grid operators are familiar with, and scaling it to a distributed system that now has millions of inverters that are capable of fast response, all of which have to operate in real-time based on system visibility (at their point of connection with the grid), knowledge that may be poor and/or delayed, and to have them continue operating even with a potential loss of communications—we can begin to understand why such a centralized system can fail, especially under disturbances and abnormal conditions.

We need a RTMR system that can operate at scale, which can benefit from the slower centralized optimization that we have today but can also ensure continued stable RTMR operation, in a decentralized and possibly suboptimal manner, when the communication and coordination breaks down, and to do it automatically. This level of flexibility at scale is largely aspirational at this time in terms of technologies on the grid but is a critical part of the functionality we need as we transition to the new future grid. The fact that we need to gradually transition from today's centralized passive grid paradigm to a distributed and decentralized active grid paradigm, and to do it while keeping the lights on, adds further complexity, uncertainty, and risk to an already challenging process.

Are Grid Operators and Utilities Ready for the Transition?

The electricity industry has clearly moved a long way from where it was in 2010 (i.e., in a state of denial that their world was changing). The utility industry today is trying to manage through the gut-wrenching change that is being thrust on it. The question now is whether the industry can adapt the existing grid to the new requirements that are becoming visible, and also, to understand what the outcome will be if they cannot adapt. They have already tried to create a forward-looking narrative by identifying the 2030 and 2050 goals for decarbonization.

In this section, we have taken those objectives and have tried to unpack and analyze the implications, and have seen major gaps in technology, investments, policy, manpower, and timeline. But because the sector is vast and technologically complex for the average nontechnical citizen to understand, and because the conversation is often controlled by the very people who are resisting disruption from these changes, there does not seem to be a coherent and holistic forward-looking narrative on what should be done. There is an implied message that now that the industry is focused on the issue, things are under control and targets will be achieved. Conversations with industry veterans confirm that it is not clear that they really know how they will get to the 2050 goals! As a utility senior executive once remarked privately, "2050 is well past my retirement window, and will likely be somebody else's problem."

Clearly, it seems unlikely that relying purely on existing institutions and directions set by incumbent utilities and grid operators will lead to the promised land. *On the other hand, the existing infrastructure will most likely still be there in 2030 and beyond, and could, in principle at least, work in tandem with the new technologies and capabilities that are being deployed to ensure a cost-effective energy transition without too much disruption in our daily lives—provided these various technologies can interoperate and work together, and regulations can be quickly modified to allow this to happen at scale.* The big challenge of course is the investment that has already been made in the existing infrastructure, the cost of transitioning to a new paradigm—both in terms of capital expenditures and new workforce skills. It is all but certain that many incumbents will resist as strongly, and sometimes as subtly as they can, with information (and sometimes disinformation) to slow down or stall the transition. This is certainly true in the developed countries where these massive infrastructure investments have been made. But what about the underdeveloped world?

It needs to be recognized that the current twentieth century paradigm has not been able to address the energy needs of over 700 million people, who still live off-grid, and over 3 billion who live with extreme energy poverty. Energy access for the people on our planet is still a big issue. This issue of energy access and equity is perhaps the best illustration of the fact that today's energy system is limited and constrained and cannot meet everyone's needs. On the other hand, these very unserved people, largely living in the "least developed countries" (LDCs) provide a greenfield opportunity to leapfrog the twentieth century grid and create a more advanced and affordable twenty-first century energy system. This issue is rarely a part of the overall discussion in energy but needs to be a major focus.

Energy Access and Equity—*The Elephant in the Room*

We have seen that in the USA, and then in the rest of the developed world, the electricity industry was regulated to ensure that everyone had (at least in principle) access to electricity at a basic level. This has become the foundational principle for most nations today—where energy is at least considered a fundamental right, even if it's not always distributed equally in practice. Left to itself, an unregulated industry would have only pursued those customers who could pay at a level that would maximize their returns and profits and would have left a large part of the nation in the dark. Public interest policy, incentives, and regulations allow these benefits to be extended to the broader population. Availability of energy is directly linked to productivity and GDP growth. The World Bank has shown a direct linkage between Human Development Index (HDI) and energy usage (Fig. 1.1, Chap. 1). Given that these data are all based on twentieth century energy technologies, it is not surprising that the wealthiest nations or those who are geopolitically fortunate and have access to petroleum reserves, enjoy the most access to energy. The least fortunate are those that use the least amount of energy, and also have to live with the lowest level of development, and concomitant HDI scores.

If we look at a satellite view of the world landmass at night, we see lights everywhere, with the exception of barren expanses of land where no one lives, and the continent of Africa. Africa is dark from this vantage because energy access is extremely limited. Over 750 million people, most of them in Africa and earning less than $1.90 per day, live off-grid with no access to electricity or the services that it can bring. Looking even more broadly, we see that there are 3 billion people who live with energy poverty so extreme that it impacts their ability to earn a livelihood.

To understand the scope of the energy access issue, we turn to Sustainable Energy for All (SE4ALL), who have defined a tiered structure for energy use—ranging from Tier 0 customers who are completely off-grid and without power, to Tier 5 customers in countries like the USA who have access to more than 10,000 watts of power for 24 h per day. Tier 1 customers have maximum power available of 1–50 watts for 4 h per day. This may be enough for a LED light or mobile phone charging, but little else. Tier 2 increases maximum power available to 100–500 watts, but still for 4 h per day. Tier 3 can now power some appliances and small machines, with peak power available of 500 watts to 2 kW, and now for 8 h per day. Tier 4 increases power level and availability to 16 h per day, increasing finally to Tier 5 as discussed above.

This problem is clearly well known and significant attention along with more than $16B/year in funding (still significantly short of what is needed) from the World Bank, IFC, IEA, and government agencies that has been focused on energy access, mainly through trying to bring the emerging market nations to a "modern" energy infrastructure. In addition, philanthropy and NGOs have played a major role in getting billions of dollars in assistance directed to improve this situation. The fundamental strategy adopted has been to bring the twentieth century grid to everyone who does not have it. This has been done at great expense, with big grants from the World Bank and loans from the IFC to nations who would deploy these technologies. Vendor companies supplying equipment and expertise have been willing partners—and in everyone's defense, maybe we did not know any better at that time. Over the last 30 years, $67 Billion have been spent on building the grid, just in Sub-Saharan African countries [7]. The thesis was that if people had access to electricity, they would consume it and their standard of living would improve, their incomes would rise, and they would become part of the prosperous global market! What actually happened is quite different!

Utilities and state actors leading grid-extension efforts in Sub-Saharan Africa and South Asia are finding little demand for electricity among newly electrified customers and, thus, little revenue and cost-recovery. Fig. 5.4 shows a World Bank chart that indicates that 95% of the utilities in Sub-Saharan Africa do not have revenues to cover their capital and operational costs, often by a lot [8].

By way of example, the average grid-connected customer in Rwanda consumed 20 kWh/month, as opposed to the 130 kWh/month they would need to consume for the utility to break even—and this story is repeated everywhere! So maybe the traditional grid is not the right answer at this stage of development—maybe we should think of microgrids and solar home systems. Microgrids can support the energy needs of smaller communities, typically using PV solar, batteries and possibly diesel generators to provide power. Microgrid entrepreneurs serve as grid operators, making the investments and providing electricity services to their customers. Given the low-level of energy consumption in low-income households, and because of the need to overrate the system to cover future anticipated needs, even microgrids have been challenged in terms of viability, and have struggled to achieve economic breakeven and scale.

At a smaller level still are solar home systems (SHS), where individual businesses or homeowners can put up a solar panel on their roof, along with batteries and power converters, to provide enough power to meet their own energy needs. Less than 1% of the $16B spent on energy access has gone into solar home systems, showing the gap in terms of buy-in from policy makers and end-users [9]. Solar home systems can be configured as low-voltage DC systems (typically 12–48 volts) powering special DC appliances, or 120/240-volt AC systems that can power traditional AC appliances, which incidentally are not designed to be extremely energy efficient.

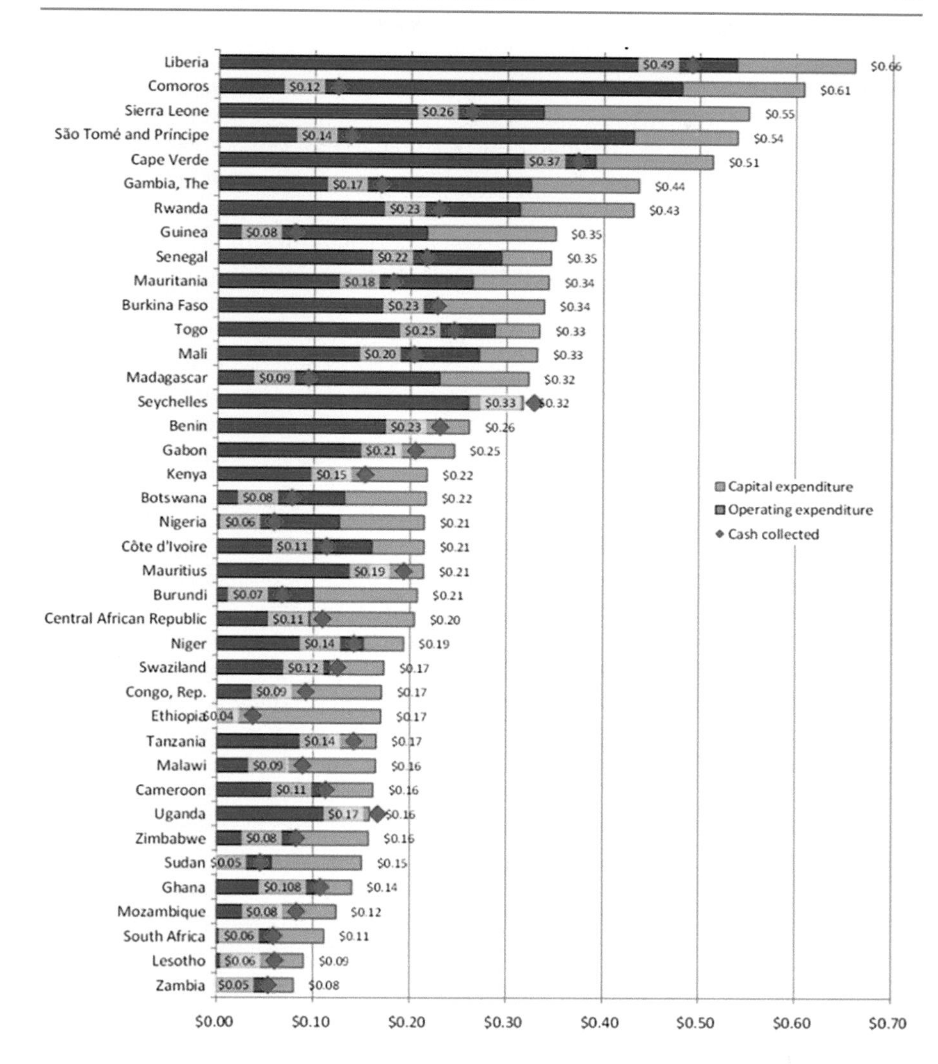

Fig. 5.4 95% of electric utilities in sub-Saharan Africa cannot cover their capital or operational costs (https://openknowledge.worldbank.org/entities/)

On the other hand, energy efficiency is important for solar power appliances, because the cost of the solar panel and inverter also have to be borne by the user. The cost of a solar home system is high—for instance a 1 kW SHS system in Nigeria costs around $1000, while a 10 kVA system, such as one we could use in the USA, can cost $15,000 to 20,000 (not affordable for the person earning $1.90/day). When these energy costs are converted to a $/kWh basis, we see that energy from a small SHS system can cost as much as $5/kWh, 50x more than we pay in the USA for electricity, and possibly 1000x more than we do in the USA on the basis of purchasing power parity. People who can afford and use the least amount of energy have to pay the most for it—this is neither just nor equitable and cannot be the basis for improving productivity and economic development or for growing energy demand.

Maybe, this requires a revisitation of our traditional views of energy access. Maybe energy access is not an "access" problem but is a "demand" problem; and may be energy poverty is not an "energy" problem, but a "poverty" problem [10]. These families need energy so they can use it for productive purposes, such as cold storage, irrigation, grain processing, machining, welding, sewing, drying, and even to operate internet cafes. These energy uses tend to increase income, and as a consequence, also increase the demand for more energy. The approach needs to be holistic and needs to address factors such as access to capital, training, energy, and markets in order for the virtuous cycle to be initiated. Providing sufficient energy for today's needs such that it is affordable today, and to then cost-effectively grow it as and when needed, is an incredibly tough problem to solve, especially with twentieth century technologies.

So, how much progress have we made? Over the last decade or so, the number of people in the world who lived with absolutely no electricity has reduced from 1.2 billion to 700 million, mostly through the availability of solar LED lighting (which sounds like and is a very good achievement!). However, during this period, only 15 million (1.2%) of these people have reached the Tier 2 level, the minimum level where the productivity and earning capacity of the family are impacted. It is clear that much needs to be done. What is even more concerning is that we are not even addressing the real energy hogs: cooking, cooling, transportation, and industrialization, all of which will increase with growth. As a result of climate change, as temperatures in parts of Asia and Africa soar, there will also be need for moderating the temperature in their homes and workplaces. To achieve all this, we will need much more than the 50 watts of power that Tier 1 offers!

This seems to be a virtually impossible problem to solve, especially viewed through the twentieth century technology lens. Most government agencies and planners are looking at long range integrated plans for building large-scale infrastructure, with poor flexibility (and high cost) using existing proven technologies and emulating the West. We now know that this will be too expensive and is unlikely to succeed. *A new approach is needed.*

Key Takeaways

In this section we have seen that as the energy transition accelerates and scales, the grid could be a critical element and enabler to make it happen. It is the backbone that connects fast-growing economical renewable energy sources with new loads, such as electric vehicles that are growing equally fast but with little to no coordination with the grid. The grid also needs to transition from a centralized passive grid to a decentralized active grid, representing a new and as yet unproven control and operating paradigm, while at the same time maintaining the reliability and cost levels that their customers have grown accustomed to. We clearly see the clash between fast-moving risk-taking commercial sectors and conservative risk-averse regulated sectors, and the many factors that govern the possible outcomes.

We also see that a major change in mindset of utility and industry leaders has occurred over the period from 2010 to 2020. The electricity industry in the USA has finally accepted that the energy transition is happening. Yet, it is not clear that, even with the best of intentions, the utility industry can transform itself in a timely manner to meet the fast-evolving requirements that are becoming clear. In the USA, we have seen some of the challenges of transforming a slow-moving regulated industry with over $3 trillion invested in the existing paradigm. We have also seen that in the Least Developed Countries (LDCs), energy access and economic opportunity are intertwined and remain a big problem, and there is a very real possibility that the LDCs will be left behind once again, even as the energy transition transforms the rest of the world.

With the complexity of so many fast-moving technologies and sectors, all intersecting through the grid (which is itself changing very slowly), the future may not only be difficult to predict, but it may also be unknowable. How do we plan for such a future, how do we make investment decisions, what actions do we take today, and how do we course correct when we have to adapt to unexpected realities on the ground? While we are still struggling to understand how these fast-moving technologies will impact us, we also need to consider whether there are other such technologies that are likely to emerge in the near future, technologies that will once again disrupt our neat view of what the future should look like. These are vexing questions, the answers to which will shape the future that we will create.

The upcoming sections explore the future scenarios that are possible and the mechanisms by which these new technologies will be translated and could scale.

References

1. National Academy of Engineering Sciences and Medicine. (2021). *The future of electric power in the United States*. National Academies Press.
2. EPRI – Research and Innovation: https://www.epri.com/thought-leadership/strategic-insights
3. The California Duck Curves. https://en.wikipedia.org/wiki/Duck_curve
4. https://www.energy.gov/sites/default/files/2023-07/Rand_Queued%20Up_2022_Tx%26Ix_Summit_061223.pdf
5. Online PV Magazine Alert. Mar 2023.
6. UNIFI Consortium – www.energy.gov/eere/solar/unifi-consortium
7. https://www.usitc.gov/publications/332/journals/electricity_investment_in_ssa-final.pdf
8. Falchetta, Giacomo, PhD Thesis. (2021). *Essays on energy and development on sub-saharan energy access, climate change and the nexus*.
9. UN Global Solutions Summit – 2023 – Divan.
10. Matt Jordan, IEEE FEPPCON Proceedings, 2019.

Further Reading

Bakke, G. (2017). *The grid, 'The fraying wires, between Americans and our energy future'*. Bloomsbury.

Gryta, T., & Mann, T. (2020). *Lights out; 'Pride, delusion and the fall of general electric*. Mariner Books.

Jones, J. (2006). *Empires of light, 'Edison, Tesla, Westinghouse, and the race to electrify the world.'*. Random House NY.

6 Getting to a Desired Energy Future: *Role of Innovation*

Rapid Changes and an Unknowable Future

Our lives have been disrupted and shaped by advances in science and our ability to access energy. This has, in turn, led to innovations and new technologies which have been developed and scaled through entrepreneurship, economics, and policy. We have seen that until the end of the nineteenth century, science often lagged innovation, with experience-based tinkering as a primary mechanism for new product inventions, building on the work of others, and finally achieving societal transformation over a period of 100+ years. In the twentieth century, scientific discovery accelerated, resulting in new disciplines that required fundamental deep domain knowledge. We saw that our mastery over internal combustion engines, electricity, nuclear power, oil exploration, and industrial controls helped to create a global economy, which along with two world wars and a cold war, also led to a new world order.

But through all this, we seemed to be in control of technology. The pace of change was relatively slow, often taking decades to centuries, allowing our nations, institutions, and processes time to anticipate and to manage the transitions. Research and technology advances were limited to a few nations and major global corporations, and the pace of knowledge generation and its transfer could be controlled. We developed planning and management processes that allowed us to take a long-term perspective, optimizing resource allocation to achieve our individual, corporate, national, and global goals, and objectives—and with great success.

However, late in the twentieth century and at the start of the twenty-first century, things began to change. We saw a few scientific advances, mainly dealing with the properties, fabrication, and control of microscopic materials, that did not seem to follow the expected patterns. These technologies showed steep and sustained learning curves, where performance improved over decades even as prices declined. At first glance, this seems similar to the traditional learning curves, for instance for steam engines, but we have seen that there are fundamental differences between the

D. Divan, S. Sharma, *ENERGY 2040*,
https://doi.org/10.1007/978-3-031-49417-8_6

two. At the same time, we have seen that as more countries developed their educational institutions, and as companies sought lower-cost manufacturing options, the basic knowledge and competencies became more universal, and in parallel, the markets and competition became global. We saw a dramatic acceleration in the timeline of introducing and scaling the new technologies, once again, resulting in legacy companies and institutions, first trying to stall adoption of the new technologies, and then trying to navigate their way through the change.

Why is it that governments, financial institutions, industry, media, and the average citizen are all struggling to understand what is going on? Why is that change is coming at us like a freight train, and the very organizations that are supposed to ensure a smooth transition, are frozen in their tracks, and are unable to take action that can mitigate the challenges and smooth the transition? Some policies that have been forward-leaning, and which understood the impact of the steep learning curves, have been wildly successful and have triggered a global tsunami of innovation and change. Yet, at the same time, the vast majority of policy makers have been more conservative in their approach and have struggled to articulate a final goal and the roadmap to get us there.

We as humans, tend to think in the short term, often using linear trajectories as we intuitively analyze possible outcomes. We have a hard time thinking of exponential trajectories, and to grasp how quickly things can spiral out of our control. When we take the exponentials and apply it across multiple interacting fields, we can see that complexity of outcomes is possibly beyond the human mind to intuitively grasp. This is true whether we are looking at the miracles of increasing bandwidth on our mobile devices, the continuing exponential growth of PV and batteries, or the exponential growth of Open AI engines such as Chat GPT.

We continue to build large complex centralized systems, often over a period of many years, because they have traditionally yielded economies of scale. But with the fast change that we are now seeing, it is becoming clear that such large systems are increasingly brittle and fragile. With rapid multifaceted change comes uncertainty, and even high-speed computation and digital twins are challenged to handle the millions of possible situations that can occur. This is particularly true in massively distributed cyber-physical systems, such as the power grid, that have a physics-based real-time-must-run (RTMR) requirement. We need to build systems that are more resilient and flexible, systems that can adapt as unexpected factors emerge and quickly become dominant, and which can operate in a future that is not only unknown but may be unknowable. But that is not how we do it today.

Ongoing Impact of Political and Geopolitical Factors

If technology and economics were the only things to worry about, life would still be very complicated. But overriding all of these are the impacts of lagging or misdirected policies and unpredictable geopolitics. In early 2022, the Russian invasion of Ukraine once again demonstrated the intertwined nature of geopolitics, oil, energy, and economics. War drove short-term shortages for oil, disrupted some of the supply chains, and resulted in extreme spikes for oil prices.

While there will clearly be ups and downs in terms of demand for coal, oil, and natural gas, the overall mid- to long-term trend is becoming clear. In many major sectors, such as transportation, electricity generation, and industrial use, conventional use of fossil fuels will likely decline over the next decade or so. This is driven by the lower (and declining) prices for solar PV, wind, energy storage, and other distributed energy resources (DERs) that are making large nuclear, coal, or gas generation plants broadly uncompetitive. This is also driven by a declaration by virtually every major automotive company in the world (along with hundreds of aspiring new players) that they are moving to all-electric fleets within the next decade. Even aviation, rail, and shipping are looking like possible candidates for this switch.

Other options being considered include CO_2-emissions-free green hydrogen for larger vehicles, or vehicles with longer daily range requirements. Based on the US pattern of energy use, 100% of coal, 70% of petroleum, and 36% of natural gas consumption will be directly impacted by this transition, resulting in shrinking markets and revenues, making it very difficult (but not impossible) for these economically challenged sectors to innovate and to transform themselves out of this disruption.

This possibly signals the beginning of a new era, where a high price of oil accelerates the adoption of lower cost renewables, which negatively impacts usage of oil, making it challenging for many of the high-cost oil producers to remain economically viable. On the other hand, when the price of oil is low, the high-cost oil producers are challenged to remain competitive—which eventually leads to an attrition in the amount of oil available. This is not surprising in markets with shrinking demand, where there is also a wide variation in cost of resource, as is the case for petroleum and natural gas. So, while we were concerned at one time about "peak oil" because we thought we would run out of oil, we think we will actually reach "peak oil" within the next 10 years because demand for oil, and not the resource itself, will decline as new lower-cost technologies take hold, and so much oil will no longer be needed. As Tom Friedman (paraphrasing Sheikh Yamani) aptly put it, "The stone age didn't end because they ran out of stones" [1]. Similarly, the oil age will end while there is still a lot of oil in the ground, and hopefully a lot less CO_2 in the atmosphere than would otherwise have been the case.

While we cannot easily predict the impact of policies and geopolitics in the near term, in many ways, the story of the diminishing influence of fossil fuels that is caused by the exponential growth and steep learning curves for batteries, PV solar, and wind is the fundamental reality that will very likely drive the energy transition in the near term—the positive climate impact of this trend will only provide a tailwind, helping to accelerate the process.

In many ways the story of the diminishing influence of fossil fuels, caused by the exponential growth and steep learning curves for batteries, PV solar, and wind, is the fundamental reality that will very likely drive the energy transition in the near term – and positive climate impact of this trend will only provide a tailwind, helping to accelerate the process.

If the price advantage for the new technologies was not there (as was the case a short 5 years ago), and this movement was to be driven purely by worries of climate change, the story would be very different: we would be facing a tremendous headwind, and this would be an almost impossible journey.

This new reality is now shaping everything including government policies, investment strategies, consumer awareness, and rapid adoption by the leading electricity and oil companies. It is also unlocking an unprecedented level of resources and investments to further facilitate the transformation.

So, the battle is won—right? What can go wrong?

Given the fast pace of change, we feel it is instructive to project possible future scenarios that could unfold if technologies keep progressing along the trajectories that we are currently seeing.

Possible Future Scenarios

We have seen that many sectors are innovating and progressing furiously, typically within their own silos. This creates tremendous progress on many fronts, but also creates the challenge of how these sectors will integrate and interact with each other and what the actual outcomes at a societal level will be. As we look forward, it is very difficult (many expert opinions notwithstanding) to predict what the future will look like, even over the next 20 years. But, as we discussed at the beginning of this chapter, a few things are becoming clear. Technologies with steep and sustained learning rates that are close to economic breakeven, which address global market opportunities, and can move at high velocity, are likely to succeed regardless of the level of opposition from incumbents who stand to be disrupted.

Technologies with steep and sustained learning rates that are close to economic breakeven, which address global market opportunities, and can move at high velocity, are likely to succeed regardless of the level of opposition from incumbents who stand to be disrupted.

On the other hand, we also have technologies that are incrementally evolving and are getting competitive with existing solutions, but which may be following more traditional linear trajectories in terms of growth, with some price declines as volume grows. The latter may appear to represent a safer investment strategy, but only if they are not disrupted halfway through the investment recovery period.

It is challenging sometimes to take a snapshot in time of various apparently competing technologies, especially some that are at an early stage, and project which one will likely ride the exponential growth curve, and which one is likely to be constrained by other intrinsic or extraneous factors. Finding these answers, and understanding the scaling constraints, and doing so as early as possible, becomes perhaps one of the important goals for understanding how to place our bets. We will briefly examine some possible scenarios that could evolve over the next 20 years, especially if growth in specific technologies continues at rates that appear likely and possible. We will also raise a flag on extraneous issues (from outside the specific technology domain) that could become important in terms of a particular scenario actually becoming reality.

Transportation

Clearly, the lion's share of the attention in transportation is being taken by the major disruption to the automotive sector that electrification seems to pose. Virtually every major automotive company has committed to electrification and is rolling out innovative and very competitive EV models. Yes, first purchase price and range may still be a problem today, but with competition so stiff and so many different battery technologies (with steep learning rates) being actively developed, it is only a matter of time before key milestones are met. Now that Lithium-Ion batteries have shown that high power and energy density batteries can result in EVs that are easy to drive and can change the way automobiles are designed, built, used, and serviced—the story has turned to address how we optimize the key element in the EV: the battery.

Battery research has focused on specific issues, such as getting more range, by using different chemistries (such as Li-Fe-PO4 and solid state batteries), different geometries (such as pouch cells), solving thermal runaway and flammability issues, managing extremely fast charging rates (so that recharge times are reduced), increasing battery life (already at 500,000 miles for some EVs), improved battery thermal management, integrating the battery pack with structural elements for safety and lower cost, reducing need for scarce materials with geographic dependencies (do not want another oil scenario), and circular economy issues including recycling of batteries with full reuse of materials such as Lithium and Cobalt. Each of these advances (when validated) can confer a competitive advantage to the innovator, and given the rapid ramp up of EV production, offers a competitive chance of getting to volume production, provided scaling issues can be fully addressed during the execution phase.

Many of the battery-cell performance issues involve science at the quantum mechanics level, requiring first an understanding of the basic principles that address the stated technical need. The solution then needs to be demonstrated at a meaningful level, and to then be scaled first to a pilot plant, and thence to a large-scale plant, managing science, technology, manufacturing, economics, and scaling issues along every step of the way. At every point, there will be new issues and challenges that were perhaps not considered in the early discovery period, most likely because they were not considered germane to the scientific question. Yet, through all that complexity, it appears that batteries have been moving along the steep learning curve that is critical for transformation of the transportation sector. This is because the key metric of $/kWh is a metric based on value (kWh directly relate to EV driving range in miles) rather than technology. This allows widely differing technologies to effectively compete for the same pot of gold.

What is left unsaid but is also an important part of the consideration is that the batteries also have to meet operational requirements, such as ability to deliver and receive peak power, weight and size that fits with automotive needs, battery life that is economically viable, and safety that ensures no fires or danger under both normal driving conditions as well as in abnormal scenarios such as severe accidents. For adoption at scale to occur, other questions need to be answered too, such as recycling, reduced dependency on scarce materials, and ability to deliver value at a societal level such that it accelerates adoption. It is anticipated that EV battery prices will continue to drop, as evinced by the frequently adjusted long-term Department of Energy (DOE) Vehicle Technology Office goal of $100/kWh and even the aspirational goal of $80/kWh, to be achieved by 2030—Ford, VW, and Tesla are hinting that even that goal can be beaten, and by a lot [2]. Given that battery prices in 2010 were at a whopping $1200/kWh, this is amazing and a testament to human ingenuity and capacity to innovate!

While EVs are dominating media coverage today, there is another major disruption brewing in transportation: electrified heavy vehicles, including vans, buses, trucks, and even large Class 8 semis. As discussed earlier, this is driven by the significantly lower operating cost for fleets of vehicles, as well as a projected lower total cost of ownership. In addition, the vehicles also save on tail pipe emissions while they are operating (whether that is a net negative in terms of overall emissions depends on the energy source they charge from). While the economic drivers are clear, the challenges are also significant. The battery pack for a large e-semi is expected to be rated at around 600–1000 kWh (with a weight of ~8000 lbs.) to achieve a 300–500-mile range. To charge such a battery in under an hour will require fast charging infrastructure rated at 1.0–1.5 MW and higher. Under today's paradigm, fleet charging of 50 trucks at a time, for instance at a warehouse, represents a peak grid demand of 75 MW needing a new substation and transmission line, which is very expensive and could take decades to plan and build, much longer than the timeline over which fleet operators are looking to deploy and scale.

While much of the discussion above has been about electric vehicles and trucks, there is another revolution underway in other forms of mobility and transportation. This includes e-bikes (cycles, scooters, and motorcycles), e-rickshaws, autonomous goods movers, railway locomotives, ships, drones, air-taxis, and electric aircraft. China and India, for instance, have launched aggressive programs for electrification of personal mobility, including the ability to swap battery packs at a diversity of locations so that an e-rickshaw or e-scooter can keep operating. Similarly, delivery of packages by drones (hopefully nothing nefarious or malevolent) is receiving clearance for use in the US and Europe. Vertical take-off and landing (VTOL) air-taxis based on quad-copter drone principles are now receiving clearance and are expected to begin operations over the next few years. Electric ships have been launched in Norway and are being tested in a variety of applications. Finally, hybrid-electric and all-electric aircraft are being actively pursued by multiple new and incumbent manufacturers—although the jury is out on whether they can successfully replace the long-haul aircraft in use today.

The other solution for heavy long-distance vehicles that require very large and heavy battery packs is the possible use of hydrogen. Hydrogen fuel cells can be operated today with an overall conversion efficiency of around 50–60%, much more than the 30% peak efficiency of internal combustion engines. One kilogram of hydrogen can be converted to ~39 kWHr of usable energy assuming all the energy can be extracted, or 16.5 kWHr with a 50% efficient fuel cell. This compares with around 12 kWHr in a kilogram of diesel and suggests that hydrogen can be an effective energy carrier. On the other hand, it should be noted that fuel cells cannot handle wide dynamic variability in output and will need a battery to source short-term peak energy requirements, which would make the overall drivetrain and vehicle more complex and expensive. There are several manufacturers (e.g., Toyota and Nikola) who are pursuing hydrogen powered vehicles. We suspect there is a future that includes both types of vehicles, at least in the near- to midterm. Whether hydrogen will succeed will depend on many factors, including sustainability, cost, safe storage, transportability, and availability of a nationwide refueling infrastructure. This is further discussed in later chapters.

While we talk about the disruption of the automotive and transportation sector, it is important to dig into this a little further. Who is actually being disrupted? At the head of the pack are the oil companies, who will lose their primary source of revenue, certainly in the case of battery-powered vehicles. In the case of hydrogen, there is an opportunity for the natural gas providers to initially develop supply of grey hydrogen, which they can further convert to blue hydrogen with sequestration of the carbon in the CO_2 produced. As we have seen, this is challenging because of cost and energy balance issues, and can only work as a reaction to a carbon tax (which seems unlikely on a global scale). In spite of their best efforts, it appears that the trend (recent oil shocks notwithstanding) is for a dramatic reduction in oil consumption as electrification of transportation continues to accelerate.

The automotive companies have declared their commitment to EVs and are making huge investments to accelerate the development and deployment of their new EV fleets. If successful, they will retain their share of the market—which is still automobiles and trucks (although newcomers such as Tesla, Rivian, Lucid, and Canoo will certainly take a significant piece of the market as well). With EVs, the incumbents stand to be more profitable as the cost of building EVs should eventually be lower than for ICE vehicles, due to simpler design and projected battery costs of under $80/kWHr over the next 3–5 years. However, it is not simply a question of putting an electric drivetrain and batteries in a conventional automobile. Companies like Tesla have changed the way that automobiles are designed and operate. Their cars are now software platforms with agile and modular code that can be upgraded with over-the-air software updates, where they can quickly try out new features on small groups of the user base to see what works and what does not.

This is contrary to the slow measured rollout of new technologies that car companies have traditionally relied on and represents a culture change that can be very difficult to adapt to and may present one of the most significant barriers facing legacy automotive manufacturers. Further, companies that have an ICE business will be conflicted because the success of EVs will be at the expense of the ICE business.

Several companies are creating completely separate divisions to avoid this conflict. On the other hand, the incumbents have deep balance sheets and the ability to manufacture at scale, and their ability to make the changes and achieve success should not be underestimated.

We do not discuss the thorny question of autonomous driving cars, especially those that can drive with complete and unrestricted access on all our roads and highways, including crowded city streets and markets. This is a very challenging problem, and may be one that will be solved soon, or maybe at a much later time. In any case, such autonomous vehicles will also require the use of electric vehicles. On a shorter timeline, we do expect to see autonomous vehicles, especially trucks, which will drive in special lanes that are restricted for the average consumer. While this is an exciting topic, one that triggers dreams of a "Jetson" like commute, it is a complex matter and success is far from assured over the next decade.

The other group that will likely be disrupted are the people that fuel and maintain the ICE vehicles. BP and Shell, amongst others, are pursuing strategies where they are trying to help existing gasoline fueling stations to adapt and to also provide EV charging. However, with the extended times needed, even for fast charging, the business models are likely to be different. EVs also require almost no normal maintenance, dramatically reducing the need for service centers that provide oil changes and routine maintenance as well as replacement of parts that wear out. So, yes, the auto industry is being disrupted by new competitors such as Tesla. New players such as battery and electric drivetrain manufacturers, as well as EV charging providers will prosper, while the traditional business of the oil companies will likely languish. New hydrogen providers are likely to emerge as strong players. In short, over the next 20 years, several incumbents will thrive while others will disappear in this normal Schumpeterian transformation.

Electricity

The other area that we have discussed extensively is the emergence of PV solar and wind as the dominant generation resources in a future grid, and the impact of their growth on the economic viability of coal and natural gas generation. We have also discussed the need for short and long duration energy storage, to balance out differences between generation and demand, both on a temporal and spatial basis. But given the dependence of renewable energy on weather, and the possibility that we can have extended periods when the sun does not shine and the wind does not blow, we also need to understand what the operating plan is for those periods where demand exceeds generation. Do we simply turn off the lights and wait, or do we have other generation resources on standby that step in and save the day? Or do we build alternate dispatchable generation that is not used 95% of the time because it is too expensive and is only used in emergencies, but which at the same time is sized to back up the entire system. Building a more expensive generation infrastructure (with poor utilization) as a backup for a lower-cost generation infrastructure, including all the transmission and distribution (T&D) to integrate it with today's system, clearly gives us the worst of both worlds and leads to substantial overbuild of infrastructure, a process that will also be very time consuming and expensive.

This can get even more challenging if we factor in the possible increase in load we have discussed. For instance, fast charging of electric vehicles and trucks can result in high peak demand, possibly with much of that occurring in urban areas and along transportation or industrial corridors. Electrification of cooking, heating, and industrial processes is increasing and will also increase electricity demand. Under the current centralized paradigm, utilities need dependable dispatchable generation if they are to meet the reliability levels that the grid provides today, a level that people are now accustomed to. We want variable green energy that is lower cost when it is available, but also want the reliability that traditional fossil generation has offered. So how do we reconcile this apparent paradox? When we further factor in our need for improved resiliency, requiring that power be available even when the bulk power system is impacted by a hurricane or fire, we don't seem to have reasonable solutions today.

Our current baseline of non-fossil resources includes nuclear and hydropower, including new variants such as small "run of the river" hydro plants [3]. Traditional large hydro plants may have a new role to play in the future—one of being able to balance the variability of renewable energy. Hydro was also normally designed to be operated as a base generation resource, capable of providing peak power for weeks and months if needed. In a departure from normal use, Norway and Denmark are using hydropower in Norway as a balancing resource for the variability of wind generation on the European mainland, using a high-voltage direct current (HVDC) link to bidirectionally control power flows to manage surplus or scarcity in renewable energy generation.

Other examples include collocating PV solar with hydropower, possibly using floating solar cell arrays, to generate power and to offset the use of the hydro resource [4]. It may seem surprising that PV can offset a mighty resource such as hydro, but some simple arithmetic shows the possibilities. For instance, the Itaipu hydroelectric project located between Brazil and Paraguay can generate 14,000 MW with a capacity factor of 62% and is one of the largest dams in the world [5]. A PV array the size of the lake behind the Itaipu dam could generate *six times* the total energy generated annually by the turbines! Given that the existing project is sunk cost, this provides one of the better integration opportunities to realize on-demand clean energy, and that too from a location where high-voltage transmission and long duration energy storage already exists (so does not have to be built). The economics are very convincing, and the fact that water can now be used for human needs and is used to generate electric power only in an emergency will also reduce the community pushback that hydro has seen in the past.

It is interesting to note that in developed countries, most of the potential hydro plants have already been built, with cost recovery achieved decades back. Simply adding PV solar to existing hydro plants can help to create an economical and clean dispatchable resource that can operate for months, even when the sun is not shining. A related opportunity to pair renewable energy with pumped hydro has seen strong resurgence, including large plants such as the new Nant de Drance plant in the Swiss

Alps, with 20 GWh of stored energy [6]. This is not only limited to the developed nations but is also being pursued in countries where the grid is being built, as seen from recent tenders for 100+ GW of pumped hydro in each of the states of Madhya Pradesh and Maharashtra in India.

In the US, we have seen that large nuclear plants have also been challenged. Existing plants are past their design life and will see an accelerating rate of shutdowns. New large nuclear plants like Vogtle have seen severe cost escalations and delays that make it very difficult to get to profitability soon and may challenge the funding of new large nuclear power plants in the US. A new approach being explored by DOE and many entrepreneurial companies is the use of small modular reactors (SMR) and microreactors. SMRs are rated at around 300 MW and can be shipped as major subassemblies from the factory and assembled at the final plant location, cutting down assembly time, cost, and risk. Microreactors are even smaller, rated at around 10 MW, and can be transported as fully operational units by semi-trailer, and can be used in remote locations, emergency situations, or even in places where space is constrained, such as in urban population centers. Regulatory approvals are in progress for both these technologies with availability projected within the next decade or so. While this looks like an interesting prospect for on-demand baseline generation, concerns of spent fuel processing remain. Additional concerns include accidents and malicious actions that can result in radioactive contamination in population centers, as well as the physical security of widely dispersed radioactive fuel, which can create additional challenges in terms of nuclear proliferation and terrorism.

In the case of transportation, we have discussed the potential for hydrogen as a fuel to power larger mobile applications. Hydrogen also holds promise in the area of power generation. While one can use more energy-efficient fuel cells for conversion as discussed earlier, it is also possible to consider using hydrogen directly in the turbines that burn natural gas today. This requires modifications to the turbines, as well as to the pipelines that today carry natural gas. Storage of hydrogen is challenging as it has poor volumetric energy density, requiring transportation and storage of large volumes of a very volatile and flammable gas. Hydrogen carriers such as ammonia and methane bond three and four atoms of hydrogen per carrier atom and have better volumetric efficiency than pure hydrogen. However, additional conversion to hydrogen through cracking or catalysis is needed prior to use of hydrogen, as well as care to ensure that emissions of CO_2 and NO_x do not occur. As discussed earlier, approaches that use existing natural gas supplies at site, but convert the natural gas to hydrogen with no CO_2 emissions and no waste byproducts are being developed and show high promise to create large amounts of low-cost carbon-free hydrogen.

There is a significant existing industrial infrastructure for producing hydrogen. Most of the hydrogen used today is made from coal or natural gas using steam methane reforming and is referred to as "gray" hydrogen. It is used for making ammonia for fertilizer and for synthetic fuels and causes significant CO_2 emissions (~22 tons of CO_2 per ton of H_2 made—accounting for total CO_2 emissions of 0.83 G-t/year). The hydrogen is referred to as "blue" hydrogen if the CO_2 is sequestered, and as "green" hydrogen if it is made from renewable energy, such as using solar energy for electrolysis of water, and has zero carbon emissions. At this time, gray hydrogen costs around $1–2/kg, with blue hydrogen costing around 50% more. Using this hydrogen for replacing natural gas for energy conversion does not make sense because the energy used to convert the methane to hydrogen creates an insurmountable cost barrier (in addition to carbon sequestration costs if blue hydrogen is used).

However, much effort is also underway to reduce the cost of green hydrogen, hydrogen that is made by the electrolysis of water using renewable resources. A key requirement is for the storage of hydrogen, which tends to be volumetrically very inefficient. However, new techniques involving composite storage tanks that can store hydrogen, one of the smallest molecules, without any leakage, are also showing promise. More will be discussed on the promising subject of hydrogen in a later chapter.

There are many large projects going on across the globe that are looking to convert inexpensive solar energy to hydrogen, including large multi gigawatt projects in Nevada, Morocco, Qatar, and Australia that are looking to make hydrogen a key part of the future clean energy system. As is the case in the middle of a fast-moving disruption, it is difficult to predict which technology will prevail. It does seem clear that the intersection of energy needs for contingency baseline generation and heavy transportation may create sufficient demand for clean hydrogen as a storable and dispatchable resource.

When coupled with low-cost PV and wind, it can also serve as long-term energy storage, helping ride the steep learning curves that PV provides. Hydrogen may also provide a better glide path to the future, as it can allow a transition phase where natural gas serves as the source for hydrogen, coupled with permanent sequestration of the carbon, either in geologic strata or in industrial feedstock. But do remember, the best technologies seldom make it to dominance and scale—there are many twists and turns along the way, and the convergence of a number of variables at the right time (aka, luck) has a big part to play in what finally succeeds. This is particularly true when the process is as chaotic as we have seen in the past, and that we see today!

The Grid—Enabler or Spoiler?

Even as we see serious efforts underway to resolve the primary energy resource issue, we also see efforts driven by entrepreneurs, investors, and companies looking to make a fortune, even as they save the world. But here we encounter the grid again—as a potential enabler of these new innovations, and also possibly as a spoiler that may inhibit, or at least constrain what can be achieved. The underlying reasons were discussed in detail earlier. What we are exploring in this chapter is what societal energy (and power) requirements will be over the next 10–20 years, and the scenario that could emerge under business as usual. Will it be incrementalism as we have experienced over the last 50+ years, or an alternative disruptive scenario that is a result of an inability of the existing grid model to cope with the change that is being thrust on it?

The incremental model that utilities are pursuing, with excellent examples out of Europe and China, as well as some US states such as California and Hawaii, are working to preserve the centralized paradigm, and to build on the massive investments that have already been made. In many of these utilities, the system is actually operating at times with high penetration of renewable energy. But as several grid operators have indicated, the stability and overall controllability of the system can be compromised, especially under extreme load, transient, or fault recovery conditions. We know we do not have a full solution for control of the future grid at this time. However, in a best-case scenario, if we assume we do have such a solution, and that it can be magically rolled out with all control issues resolved (even at high IBR penetration levels) at some time in the future using technology that is currently being developed, can we achieve our overall goal?

In such a best-case scenario, we have to further assume that the utilities (especially in the US) can build T&D infrastructure to accommodate PV solar, wind, energy storage, and other new clean energy sources that can bring the "lower cost" renewable energy into the major urban and peri-urban load centers. This has to be done while ensuring reliable power even when renewable energy is not there. They also need to supply ever faster rates of EV, e-semi, and other e-transportation charging, without overloading their transmission/distribution lines and transformers. As commercial and industrial loads electrify, they also need to meet the new demand. They also need to coordinate the power generated by millions of residential rooftop inverters with grid needs. The bulk system (and the distribution system) also presents a major target for hackers and cybercrime and needs to be protected. The new system also needs to be resilient, ensuring that the grid edge has power when the bulk system is impacted by a high-impact-low-frequency (HILF) event. And oh yes, this all has to be done over the next 10–20 years and at a much faster pace than utilities are used to, and on a budget that regulators will approve of!

If this can all be done in a timely manner, there is a chance that the utilities can get to their stated goal of 0% carbon emissions by 2050. This is a major challenge, especially in a country like the US, with fragmented and fractured regulations and dead-locked politics that make sweeping major change very difficult, if not impossible. *But we sincerely hope this is feasible and works the way the utilities would like—it will certainly be the preferred approach, as it will minimize disruption for utility customers and for the utilities as well.*

But as we have discussed, the grid edge is not waiting for anyone. PV solar, wind, energy storage, EVs, data centers, hydrogen, electrification are all continuing their relentless and unwavering march forward—all with an assumption that any energy that is generated will be used as and when the generator can provide it, and that the energy required by loads will be available whenever and wherever they want it. In the past, because customers were highly dependent on the grid for power, a lack of access to the grid would have been a showstopper.

Today, when self-generated energy is at or better than grid parity in many cases, and the cost to the commercial entity of not rolling out the EV fleet or the datacenters is too high, the equation changes. If we look at large datacenters, or at logistics centers that will eventually serve as transportation hubs, they understand the challenges associated with getting grid services in a timely manner. Many of these new facilities are being built with captive generation, which in fact also have the capability to provide grid services and support. If the generation is clean and includes storage, we are able to (at sufficient scale) realize a cost of power that is competitive with the grid.

A similar argument is playing out with commercial and residential customers, where they are putting in PV solar and storage to offset their cost of energy and to meet their internal "green" targets, as well as to have backup power available for emergencies. When high peak demand loads (such as EV fast chargers) are to be served, the cost of the local generation can be justified on the basis of the demand charges that have to be paid. As the cost of PV and solar decline even further, and hydrogen becomes cost competitive, this trend could become even more prevalent.

This suggests that over the next 10–20 years, if the utilities cannot significantly expand their T&D networks, a very large amount of generation (800 GW+) could be built at the grid edge and along the distribution network, mainly to power these new loads, and to ensure that is done in a timely manner. Figure 6.1 shows the expected increase in bulk generation that is expected, as well as the increase in new load. If adequate new transmission capacity is not built, it will force the deployment of new generation at the grid edge.

This new grid-edge generation could be beneficial in the sense that the utility has to build less infrastructure, but also leads to more challenges in managing the grid and integrating these new customer-owned generation resources into the grid. This can also lead to reduced demand on the grid due to self-generation by customers, which in turn increases the danger of customer defection, especially if the utility is unable to provide service with high reliability and low cost. *In many ways, this would be a worst-case scenario because only the customers who cannot afford to*

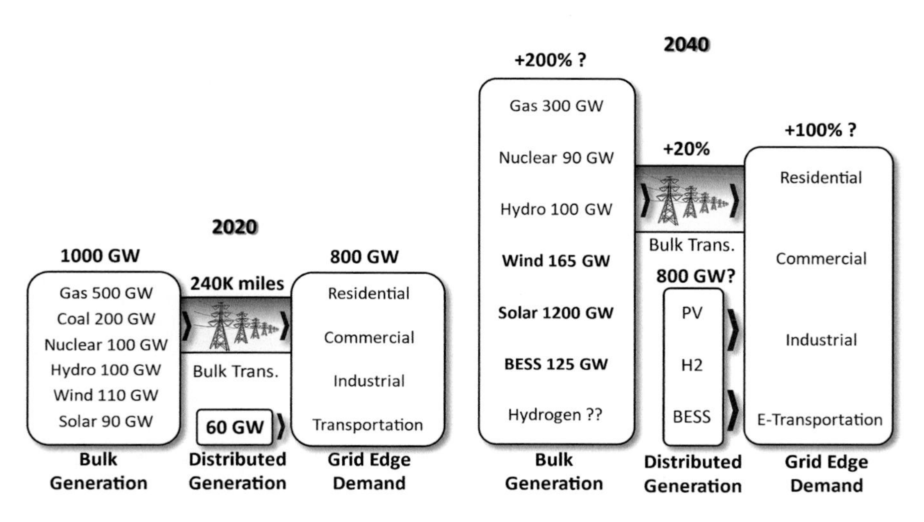

Fig. 6.1 Challenges with building transmission in the US could force developers to build generation capacity at the grid edge close to where the new loads are sited

self-generate would remain, reducing demand and revenues, and creating the danger of a death spiral for the utilities. Under the proper mix of regulations and incentives, generation at the grid edge could be used to provide grid services and resiliency for the entire grid, especially in the hard-to-serve urban areas.

Looking at all these issues, it appears that a number of possible scenarios could emerge in terms of how the grid evolves over the next 10–20 years.

Scenario 1 - Centralized Grid In the first scenario, as discussed above, the grid can continue to evolve the way the utilities and regulators are planning. This could result in a substantial overbuilding of new infrastructure and would be very expensive but would generally preserve the centralized paradigm. It is also not clear that, given regulatory delays and a fragmented regulatory framework, the needed infrastructure build could be done in the time available.

Scenario 2 - Decentralized and Fragmented Grid A second scenario is more dire, where decreasing reliability and higher costs on the grid drive customers toward self-generation. Giving in to public pressure, the regulators allow customers to band together to create microgrids and minigrids (possibly extending the existing co-op model) that improve reliability and reduce their dependence on the bulk grid. In this Wild-West scenario, many utilities may not survive, raising the question of who then will provide the public infrastructure for industry and to service the economically disadvantaged.

Scenario 3 - Integrated Grid A third possible scenario is a mashup of the two previous scenarios. Utilities use their bulk assets to deliver low-cost power most of the time. Distributed local self-generation and storage is allowed to increase and is paid for by the end-user but can also be dispatched by the utility to provide peak demand management for loads such as EVs, as well as for providing grid services and community resiliency support when needed.

Vertically integrated utilities can, with regulator buy-in, take ownership (or control) of the distributed generation source, for instance a hybrid (PV plus storage) community farm or warehouse rooftop plant, and provide some of the benefits to the owner (such as backup power and lower utility bills), and use the distributed resource to help manage the grid-edge. For utilities that operate in ISO markets, they can still build the generation plant, but can auction/sell off the PV panel (the actual generation source—the rest is simply power processing gear which the utility can own), passing the cost on to private buyers, providing them returns on their investment based on the energy generated [7].

However, because the utilities can operate the plant as needed for supporting grid operations, their integration costs would be low, and they could rapidly scale this new resource. By having control over distributed generation resources, utilities can also significantly reduce the need to build large new generation plants and the new T&D infrastructure to bring the power in. As the NASEM report on the future grid suggested, reliability and resiliency will possibly come from the grid-edge, while access to low-cost power will come from more economical bulk resources (which will be most of the time) [8].

Scenario 4 - Dystopian Grid Finally, a fourth scenario, unfortunately, is also possible. PV solar, batteries and EVs continue their downward march in prices, but utilities and regulators, leery of disruption of existing operating models and concerned about risks and a lack of standards for these new emerging technologies, slow down adoption dramatically through draconian grid integration studies, expensive interconnection processes, and long-term pilots (remember "death by pilot"). Media and social media continue to dramatically amplify every event—from a blackout (whether caused by climate change or equipment failure); hiccup in the PV, battery, or EV supply chain; teething problems such as isolated battery fires or wind turbine failures; or gyrating oil prices caused by geopolitical issues—positioning all these events as a complete failure of the new technologies, and a cautionary tale about what happens when we move away from proven solutions. Shifting political winds result in gyrating inconsistent strategies that constrain the ability to make progress, or to even look at issues with anything other than a two- or four-year election cycle window. *It's a hell of a way to manage the world—where a single uninformed Tweet can have more power than a climate agreement that was refined over more than two decades by the world's most qualified scientists and policy makers and signed by 193 countries plus the European Union*. It should be clear that a chaotic process to decision-making can result in an equally chaotic end result.

It should be clear that a chaotic process to decision-making can result in an equally chaotic end result.

In such a Orwellian scenario, we could wake up in 2040 to a world where solar energy is stalled, EVs are unable to fast charge, electronic and battery waste is exploding, coal and nuclear plants have been reactivated, energy is more expensive, distribution networks are overloaded with the utilities valiantly trying to keep a tired and dated infrastructure operating, and grid reliability is poorer than we have it today—and oh, by the way, the impact of climate change is also upon us! Islanded microgrids are everywhere, supplying power to those who can afford it, insulating them from the unprecedented heat waves and cold snaps and helping them adapt to the impacts of climate change, while the economically disadvantaged live in a dystopian world where their misery is amplified by poor access to energy and related services!

Which scenario will actually evolve remains unknown and would largely be as a result of our own actions over the next decade or so.

Can We Get to Universal Energy Access?

If we could have a clean slate to build a future energy system, what would we wish for? We are seeing that modifying the existing infrastructure poses some problems. Maybe, the greenfield opportunities in the least developed countries (LDCs), where there is no grid infrastructure and most of the people live off-grid, can provide a template for how the new infrastructure should be built. Can the twenty-first-century exponential technologies make a difference in those cases?

We have seen the power of rapidly evolving technologies and steep learning curves. One approach could be to build the future system using modular plug-and-play building blocks that use twenty-first-century technologies with steep learning curves and which can be manufactured in high volume and easily assembled at site, can be interconnected to increase system capacity in an ad hoc manner as needed, are interoperable across many vendors and technology generations, can be installed, operated, and maintained by local staff with low expertise, can connect and operate with the existing grid when required, can manage billing and transactive issues as needed, and are infinitely extendible! None of these attributes are available in existing solutions, but this is all feasible today using some of the technologies described in earlier chapters. *What is needed is a forward-looking policy that incentivizes the deployment of such devices and uses results-based financing to ensure that societal objectives are achieved.*

We discussed how Germany's Feed-In Tariff (FIT) program was a trigger that drove down PV solar prices at a very rapid pace. Similarly, India's "Unnat Jyoti by Affordable LEDs for All" (UJALA—means light or radiance in Hindi) established a competitive bidding process for sourcing LEDs for the Indian market and helped drive LED prices down from Rs. 400 in 2014 to Rs. 70 in 2018 (around $1), which in turn drove the volume of installed LEDs in India from five million in 2014 to 669 million by 2018, a short 5 years later. As a direct result of this transition, India is saving over 30 TWHrs of energy per year, equivalent to Denmark's annual electricity usage, which with today's fossil-based generation, also results in a dramatic reduction in carbon emissions. This is another classic example of a forward-leaning holistic policy driving price declines along the steep learning curve for LEDs. Clearly, checks and balances are needed to ensure quality and compliance with overall objectives, which can also be very challenging.

Such an approach can provide an opportunity for those countries looking to leapfrog twentieth-century electricity infrastructure, and move directly to a modern twenty-first-century grid, that is more sustainable, affordable, and can be expanded as needed. Over time, the full capacity of a modern grid, with high capacity and reliability and low cost, can be achieved, but it allows them to get started at the right level and the right cost. By using systems that are modular and interoperable through architecture and design, system growth can occur unimpeded at the rate that is needed, even as technology continues to change rapidly. But these attributes of technology and vendor agnostic interoperability will not happen automatically but have to be driven by policy and standards. It is interesting to note that this is a somewhat similar pathway as when China embarked on their plan for grid modernization and upgrades. With an integrated objective of deploying the new grid and simultaneously developing self-reliance in all aspects of grid solutions, they also followed a trajectory where the solutions they deployed were often initially at a much lower sophistication and reliability level than was available from global vendors such as ABB and Schneider (who they had often co-opted the designs from anyway). But with market experience, a tolerance for early failures, and rapid design iterations, they were quickly able to reach and surpass global reliability and performance standards.

Can such a leapfrog happen by 2040? In today's world, two decades is a very long time—think of the period from 2000 to 2020! The fact that PV solar and storage, possible bedrock technologies for the early stages of this future grid concept, are already at better than grid parity, and are continuing steep declines in price along with continuing performance improvements, makes this a relatively low-risk strategy. Further, we are already spending $16B/year on rolling out an outdated twentieth-century grid in the emerging LDC nations. Even a portion of that funding over 10+ years can exceed $100B in new investments and can provide a very substantial incentive for manufacturers: to design the right products; to get the basic infrastructure; and to kickstart economic development for the target communities.

However, national integrated energy plans generally rely on existing solutions and cannot respond to fast moving technology cycles. Examples such as UJALA are outliers and are designed to accelerate the fast technology cycles even further, and to use the concomitant price/performance benefits to achieve overall societal goals. Success will require a simultaneous top-down and bottom-up implementation, a shared view of success which should include economic development and not simply energy access, incentives to ensure that organizational silos do not inhibit progress, and directed funding and incentives to achieve holistic goals, including sustainability, flexibility, equity, and a circular economy. It is important to emphasize that energy access is not a silver bullet but should be viewed through a broader cross-sectoral lens including food security, access to water, health and economic development [9]. As we will see in later chapters, we are at a unique point in time where such broad progress is possible.

This new future grid concept is also of interest as global conflict grows and even more refugees are added to the 82 million already displaced from their own countries. Access to energy becomes a key enabler for them as well. As climate change drives an increasing number of costly HILF events, typically incurring losses of more than $1 billion or more, the question of resiliency and rapid power restoration for impacted communities becomes an important consideration. A bottom-up grid of the type described above could substantially improve community resiliency. Finally, the emerging markets also offer the developed world a unique opportunity to build and test the ideas that underlie their own future grid. Risk of poor performance is acceptable here in the early days, as the baseline is no power at all! Once the basic ideas have been validated, the technology can gradually upgrade today's centralized grid, possibly with something more decentralized.

Energy in 2040—Utopia or Dystopia?

The next two decades could be very consequential to the way we live our lives, the comforts we enjoy, and the type of world we will leave behind for our children and grandchildren. Without abundant, economical, and reliable energy, the world will definitely be dystopian. On the other hand, with abundant energy, we can solve a lot of the issues and problems that we have discussed above. At the beginning of the twenty-first century, there was no viable path to a future that did not depend completely on fossil fuels. The geopolitical dependencies of oil divided nations and their people in terms of haves and have-nots. Endemic energy shortages, and an inability to deal with climate-related issues were visible to us all, but we were unable to do anything about it—truly a tragedy of the horizon.

Twenty years later, we have shown that a bright future is possible. Technologies exist which can enable an era of abundant, economical, reliable, and resilient energy. We see an opportunity to solve the grand challenge of energy access in emerging economies, with the same technologies that can address the energy needs of the developed economies. What is particularly exciting about this time around is that for the first time, we are not fighting a zero-sum game, where someone wins only when someone else loses. Promising new technologies perform better than existing solutions *and* cost less. We are also seeing technologies that seem to break the hegemony of bulk materials, allowing steep and sustained learning curves where the price of a product or service declines even as functionality and performance improves. But these new technologies are often in direct conflict with the entrenched incumbents who stand to be heavily disrupted, with staggering economic losses for them if they cannot develop and execute on a transition strategy. Further, lagging or politically driven policies can dramatically impact the ability of the new technologies to take root, causing more chaos and confusion. It is the interaction and intersection of these many forces that will determine which of the scenarios will actually play out—and the outcomes may depend on politics and region. As a result, which scenario actually develops will depend on the choices we make.

At the same time, if we let industry drive the expansion of these fast-growing segments, we are likely to see a narrow and siloed view of the problems—driven by competitive pressures, need for sales and profitability, as well as to manage funding and finance. Left to itself, industry would focus on availability, price, and market dominance, and we could easily get into a situation where Lithium or rare-earth metals become the next oil, and we again become geopolitical hostages.

Forward-leaning policies and regulations can drive a more holistic approach, providing incentives to accelerate time to breakeven for technologies that are beneficial at a broader societal level, while ensuring that market expansion issues such as interoperability are also addressed. This should also include issues such as circular economy, equity, access, preferred use of widely available resources, and sustainability. This requires attention to many issues that are not front and center for the companies doing PV solar, battery and electric vehicles, and involves the integration of additional new elements that can be viewed as a distraction.

As we have tried to understand what is driving these fast-moving segments, it has become clear that these are the exponential technologies with steep learning curves that have risen from an improved understanding of quantum physics, and our ability to manufacture objects with microscopic and even nano scale features and properties in high volume. But even with these key technologies, we have also seen a 50–70-year translation cycle from science to impact, time that we do not have. A related question is whether these are the only scientific discoveries and technologies that can be brought to market over the next 20 years. History suggests that is the case. However, looking at the process of translation from lab to market, we also see that with the disappearance of the corporate long-term R&D laboratory, the science-to-market process itself is broken. Does that mean the translation will slow down even more?

If we could accelerate the time to market for promising scientific discoveries and technologies, such as green hydrogen, permanent removal of atmospheric CO_2, and access to pure water, and if we could ensure that the outcomes were aligned with broader societal goals and objectives—we could move toward a utopian view of our future world.

If we could accelerate the time to market for promising scientific discoveries and technologies, such as green hydrogen, permanent removal of atmospheric CO_2, and access to pure water, and if we could ensure that the outcomes were aligned with broader societal goals and objectives—we could move towards a utopian view of our future world.

But who is going to do the work needed to achieve this acceleration? The roles of academia, research labs, and industry have shifted substantially over the last 30–40 years, with industry abandoning their role in fundamental research and translation. In the energy field, utilities do not do active research or maintain skills to drive the transition to a new power electronics-rich system. This makes a utility-led transformation of the existing power grid very challenging and unlikely. So how can we fill this major gap?

Can We Get the New Transformative Technologies We Need by 2040?

As we have looked at the likely scenarios that could unfold over the next 20 years, we have seen that there are several major factors underlying the disruptions and transformations that are underway. True, we are already seeing the impact of climate change in our day-to-day lives and are also seeing an attempt by many countries to formulate a plan to mitigate the impact in the long term, and to adapt to its unstoppable force in the near- to midterm. In the near term, upheavals caused by global epidemics, wars, recessions, and inflation get in the way of consistent actions, and demand expedient political responses that mitigate the pain for the people in the short term—all of which is understandable and necessary. *However, climate change denial that is rooted in politics and not in scientific evidence and facts does not contribute to the discussions in any positive way except to create a divided community at a time when we all need to work together.*

The big impact of climate change that will force people to take action is driven by the economic cost and related consequences that individuals, communities, and countries are beginning to face. Insurers who are repeatedly having to cover the cost of billion-dollar hundred-year events (HILFs), that now seem to occur every few years, are raising insurance rates for susceptible properties, activities, and businesses, causing homeowners and companies to take notice. Activist shareholders are driving corporations to adopt sustainability goals and are pressuring them to move toward green energy use. Policies and incentives (often on a state-by-state basis in the US) are also creating opportunities and rapidly expanding new markets for entire new industries that simply did not exist as recently as 2000.

We are finally at a point in time when many of the needed technologies exist and the economics are favorable for us to take action on the climate front—something that has eluded us for the last 20+ years. But none of this would be possible without the new technologies that underlie the disruptions in the field of energy. As we have seen, the accelerated pace of change that we are experiencing is coming from just a few new twenty-first-century exponential technologies with steep and sustained learning curves (e.g., PV solar, batteries and semiconductors), where performance improvements coupled with price declines and global markets have created a massive wave of change in a very short time, and at a pace that is unprecedented in our history. Many of these technologies are based on a new and deeper understanding of materials and their properties, and the ability to manipulate materials, movement, and energy flows, and to do so at rapidly declining cost. This has required science, innovation, engineers, entrepreneurs, investors, and aligned policy. Massive investments have been made to exploit the market opportunity that these disruptions present. And if the past is any guide, things will take 30–50 years to settle down to a new normal as the chaotic transformation process is completed. While that was the only option in the past, this time around, the delay will exact a steep penalty from all of us because it will be costly and chaotic and will limit and delay our ability to address climate-change-related impacts and to rectify the damage we have caused to our own world.

We need to find a way to accelerate this transformation. For this, we first need to revisit how the process of going from Ideas-to-Innovation-to-Technology-to-Market-to-Impact works.

The end of World War II saw a fundamental shift in the way the US, and eventually the world, did scientific research. It was becoming clear that major scientific advances (as opposed to tinkering) underpinned many of the new technologies and capabilities that were transforming our world, both in terms of national defense and growth of the economy. Funding for scientific research grew dramatically, and major research universities as well as industrial and national labs, began a furious process of scientific discovery that has continued unabated. In the mid-twentieth century, turning these scientific discoveries into technology and societal value was not in the realm of academic research (in which we include our national labs), but of industry. Companies such as General Dynamics, General Electric, IBM, Bell Labs, and Westinghouse had long-term R&D programs (with 10+ year horizons) which could systematically analyze new innovations (including for a lot of research they themselves had done), and to then create a roadmap along which these technologies would be commercialized. Without much competition, this was the safe and responsible way to get science converted to value and impact.

At the end of the twentieth century, coinciding very much with the end of the Cold War, we saw the emergence of two parallel threads with massive impact. Firstly, we saw a significant increase in the pace of change in scientific discovery and its rapid adoption in our day-to-day lives. This includes digital devices, communications, and the internet. We also saw the emergence of new international players who were able to rapidly commoditize many of the twentieth-century technologies, thus increasing competition for consumer products. Secondly, at the same time, and for similar reasons of maintaining global competitiveness, major incumbent industrial players in the US abandoned long-term research, and shortened their time horizon to commercialize innovation, first to 3–5 years, and eventually to 1–2 years, looking at product development and commercial deployment as their key goals.

So, while universities and national labs were doing groundbreaking scientific research, who was doing the conversion of that scientific research into value?

Universities were singularly unprepared for this task. Entrepreneurs were keen to step into the void but needed funding. Venture capital was open to step into the gap but needed certainty that the technology would work and that the major "showstoppers" had been identified—which the scientific research did not provide. It was left to a rag-tag band of entrepreneurs and angel-investors to try to piece this puzzle together—matching a specific scientific discovery to its potential value and possible impact in the market, and to then reducing risk so that this could become a viable company.

This is extremely challenging because the scientific R&D team typically looks at a problem very specifically, typically to address a narrow scientific feasibility question, and often looks at a publication in a reputed journal as the desirable end goal—not taking the product or technology to market.

As an example of a science question that can have very high impact, let us look at the potential for extraction of Lithium from seawater. Science today shows that this can be done—which is a great first step. But hundreds of questions remain. Can this be replicated and scaled? Do we understand the process and its sensitivities? Are there limitations that would make this prohibitively expensive? If this were successful, how would it compare with current Lithium extraction techniques? In terms of a circular economy and waste stream management, are there any challenges? If the technology could be successfully demonstrated, how could this get to market? How many years would this take, and what level of investment, team, and facilities would be needed?

Not only is this spate of questions very challenging for the scientific researchers to answer, but it is also very difficult for the entrepreneur and team to answer, because a big part of the deep-domain knowledge in the field that the discovery was based on is lost as the technology moves out of the lab and to a startup. Because scientific papers and reports all talk of successes, the challenges and issues that can deter success are often not talked about or are glossed over—which makes it difficult for the team exploring commercialization to have confidence that key challenges and innovation risks, that are a major concern for them, have been understood or addressed.

Because scientific papers and reports all talk of successes, the challenges and issues that can deter success are often not talked about or are glossed over—which makes it difficult for the team exploring commercialization to have confidence that key challenges and innovation risks, that are a major concern for them, have been understood or addressed.

Further, there is a lot of confusion amongst scientific researchers about the entrepreneurial process itself and the steps needed to create an operating company, especially for companies based on deep-domain knowledge and transformative science that can disrupt existing business models and paradigms. Standard cook-book concepts such as customer-discovery and voice-of-customer are more applicable for companies seeking to exploit incremental innovations that provide competitive advantage in existing markets, and which can then be compared in terms of functionality and value to existing solutions in the market. Henry Ford seems to have captured this well when he is reputed to have said, "If I had asked people what they wanted, they would have said 'faster horses'!" [10]. Market incumbents and users of current technologies often cannot visualize how a new nascent technology will impact them or can grow to become a market leader—examples abound including personal computers, mobile phones, solar cells, and electric vehicles! It is not surprising that translation of science to impact is not a smooth process, and that success rates for these companies are extremely low (as will be covered in detail in the next chapter).

In spite of this challenging process of tech-to-market (T2M—as it is often called), today we are sitting on the cusp of a major transformation in virtually every aspect of our lives. In the field of energy, we have generally boiled it down to the impact of specific technologies with steep learning rates, including PV solar, advanced batteries, and semiconductors. But it has taken us 70 years to get from first principles to a point where we have achieved price parity and the market impact is real. Now that the larger companies have eliminated or substantially scaled back long-term research, will the next disruption take even longer, and will it be more inefficient and chaotic?

Today, there are hundreds of scientific discoveries, all of which are built on fundamental knowledge that was not known even 10 years ago, and certainly are in technology areas where the common man has little by way of intuition. These discoveries can potentially be transformative for our future—including potential outcomes such as green hydrogen generation, quantum dot, and organic solar cells; CO_2 to cement and fuel, pure water, and rare minerals from sea water; nanometer integrated circuits; quantum computing; and so on. These new discoveries are generally sitting in a university or research lab somewhere around the world or have been published in a technical paper or PhD thesis where the application may not be clearly defined.

Even if someone was very interested in finding a solution that required, say Lithium extraction from seawater, there are likely to be hundreds of papers that discuss potentially relevant scientific principles that could be applied. As an interested industry partner or entrepreneur, how do you figure out which one is the best approach? The people who have the most knowledge have checked out, the PhD student has graduated, and the professor has moved on to another even more exciting project. In any case, most academics get no brownie points for the mundane and very difficult task of taking their new ideas all the way to commercialization—so the technology license is often an orphan—sitting there as a patent document, PhD thesis, or technical paper, with little additional information to reduce risk or accelerate commercial development.

As we have said, in the past there may have been no other option to this random and disorganized *process* (for lack of a better word), and it took decades for new scientific discoveries and disruptive technologies to get to market. There is a need for a process by which these new potentially high impact scientific discoveries can be accelerated to commercialization. The US government recognizes this issue and has started programs such as NSF I-Corps, which was created in 2011 to help move academic research to market. Similarly, ARPA-E and DOE have a strong T2M focus and require funded research teams to show translation of their work toward commercialization. The Technology Readiness Level or TRL metric has been created to provide guidance on how close to commercial readiness a particular technology is. These factors certainly provide an indicator that people agree that there is high value in the research, and that effort is needed to bring it out. But success as measured by actual societal impact has been very limited for some of the reasons discussed above. The next chapter will examine the journey from science to impact at scale, the challenges, and new possibilities for accelerating science to impact.

We hope that our collective societal "WE" will have the wisdom to understand the gravity of the current situation, and the importance of this unique point in time where both economic and societal objectives are aligned for the first time ever, so that we might be able to take advantage of the moment. At the same time, we worry about the tragedy of the horizon and our inability to take action. We are hopeful that our self-interest, rather than the perceived virtues, will guide us and allow us to identify and select the midterm economic benefits that we can get for ourselves and for our families and use that to drive us, guided by the "guard-rails" of policy, to a better scenario than Scenario 4 discussed above.

The above discussion simply shows that the future energy infrastructure will be determined by many fast-moving technologies and market segments, all of which will interact with each other and with other slow-moving segments in unpredictable ways, even over a short span of 10–20 years. What actually transpires will depend on policies, geopolitics, incentives, economics, but more importantly also on scientists, engineers, innovators, entrepreneurs and companies (both small and large), and VCs and financial investors, *and how quickly they can translate new science into impact at scale*. We have seen that in the US and other developed countries, the huge investment in both the petroleum and electricity infrastructure can create reluctance to drive change, because of financial losses and uncertainty and chaos that can come with major transitions and change. Yet the change now seems unstoppable. We have reached the tipping point where these new better performing solutions are often lower cost than the existing solutions, are riding steep learning curves where the cost goes down even as performance improves, and where we finally do not have to compromise between climate and family, between wallet and planet!

We have reached the tipping point where these new better performing solutions are often lower cost than the existing solutions, are riding steep learning curves where the cost goes down even as performance improves, and where we finally do not have to compromise between climate and family, between wallet and planet!

But what about emerging economies and the three billion people who live off-grid or in extreme energy poverty all over the world in places where the energy infrastructure is not yet developed. Do these economies have a chance to leapfrog from nothing to a modern sustainable energy infrastructure in one step? Or will they be left behind once again, trying to catch up to technologies that are even more sophisticated and are moving faster than before? *Is there a green field opportunity to build a new energy infrastructure that is both just and equitable, as well as affordable and decarbonized, and provides an opportunity for these countries to leapfrog the developed world?* This is a critical issue that needs to be addressed as we look for twenty-first-century energy solutions, to both meet our economic growth needs, as well as to mitigate the impact of climate change.

We see the level of upheaval and disruption that has been caused by only a few select technologies, where these technologies have shown great potential in terms of solving some of our most pressing problems. But they also raise the question—are there other such high impact technologies that are available in research labs today, and if those could have an impact by 2040? The traditional view is that any adoption at scale would be too slow, but the last decade has shown us other possibilities, particularly when exponential technologies are involved.

But who is going to do the work needed to achieve the needed acceleration? The roles of academia, research labs, and industry have shifted substantially over the last 30–40 years, with industry abandoning their role in fundamental research and translation. In the energy field, utilities do not do active research or maintain skills in the new emerging areas and are unlikely to drive the transformation. So how can we fill this major gap?

We feel that the time is right to reassess the role of major research universities and laboratories (the place where the deep domain expertise exists) in this technology development and translation process. We need to separate out the pathways for new ventures based on incremental innovation around principles that are well understood in the market, and disruptive startups that have to carve their own path. Key questions to address are: who are the partners along the way; how does the technology transition out of the lab; how is it de-risked so that investors feel comfortable to move forward; and how does it get to scale? Are there new institutions and processes needed, or can we manage with what we have? These are questions that challenge the present research, innovation, tech transfer, and commercialization process. The next chapter looks at how we can accelerate the time from science to impact, at scale.

References

1. Original phrase attributed to Ahmed Zaki Yamani (Minister of Oil of Saudi Arabia) and as discussed and referred in public domain by many including mentioned by Friedman, T. (2000, September 8). The secret oil talks. *New York Times*. 'Hot, Flat & Crowded', 2008.
2. Reuters Report (2023, September 7): https://www.reuters.com/technology/ev-energy-storage-battery-prices-set-fall-more-report-2023-09-07/
3. British Hydropower Association. (n.d.). *Small run of river*. BHA. https://www.british-hydro.org/small-run-of-river/ (Using the natural flow of rivers, small-scale hydro is a highly efficient form of energy).
4. Itaipu – Largest Hydroelectric Dam (Brazil/Paraguay border): https://en.wikipedia.org/wiki/Itaipu_Dam
5. 'Energy 2030' Conference at Georgia Tech – Presentation by Divan (2008).
6. Nant de Drance, Switzerland: https://en.wikipedia.org/wiki/Nant_de_Drance_Hydropower_Plant
7. Paul Centolella – DOE Project Reports and Publications accessible through: https://www.linkedin.com/in/paul-centolella-b3ab7550/
8. The Future of Electric Power in the United States (2021). The National Academies Press. http://nap.edu/25968
9. Empower Billion Lives – Global Energy Access Forum: https://www.ieee-pels.org/programs-projects/geaf/ebl
10. Henry Ford – Autobiography references: https://en.wikipedia.org/wiki/Henry_Ford, and several references pointing to lack of inadequate evidence whether Mr. Ford really said. However, we chose to quote this reference in the context of innovation.

Further Reading

Chesbrough, H. W. (2006). *Open innovation*. HBS Press.

Christensen, C. (2012). The innovator's dilemma. *Harvard Business Review Press.*

Denson, R., & Kanter, M. (2015). *Advancing a jobs-driven economy – Higher education and business partnerships lead the way*. STEM Connector. Morgan James, The Entrepreneurial Publisher.

Drucker, P. F. (2001). "*Innovation and entrepreneurship*" first published in 1956, and later gain in 1992 by Collins. Also, "*The essential Drucker*" by Harper Business.

"Frontiers of Engineering" – National Academy of Engineering – Reports of Leading-edge Technologies Impacting Upcoming Decades (2015).
Panicker, N. R. (2021). *An entrepreneur's journey – The joy of dreaming*. Notion Press.
Pelton, J. N., & Singh, I. B. (2015). *Digital defense*. Springer.
Sharma, S. K. (2014). *Energy: India, China, America and rest of the world, Chapter 8; The 3rd American Dream*. Creative Publications, Amazon.
Sharma, S. K., & Meyer, K. E. (2019). *Industrializing innovation – the next frontier*. Springer Nature.
Shrier, D., & Pentland, A. (2016). *Frontiers of financial technology*. Visionary Future.
Siota, J. (2018). *Commercializing discoveries at research centers – linked innovation*. Palgrave Pivot MacMillan.
Thomas, H., Lorange, P., & Sheth, J. (2013). *The business school in the 21st century*. Cambridge University Press.
Yergin, D. (2011). *The quest: Energy, security, and the remaking of the modern world*. Penguin.

Accelerating Commercialization of Energy Innovations

7

Every dark night awaits a dawn. A quiet new revolution in our pragmatic ability to successfully commercialize innovations for scale, speed and impact is beginning to show promise, and can dramatically impact ENERGY 2040.

—Authors

Introduction

The complex mosaic created by the rapid energy transition, now underway for the past 20 years, makes it difficult to predict how the future will actually unfold. Improving our ability and understanding of how-to bring disruptive energy innovations-to-market in a speedy, scalable, successful, and economically viable way can improve both our ability to predict trajectories, as well as the actual outcomes. However, to understand how the "innovation ecosystem" needs to change, we need to first understand where we currently stand.

In this chapter, we take a deeper dive to recap how and why the process of bringing energy innovations to market is so inefficient, and why it has evolved so slowly over the last century. It has also been surprising to see, existing successes notwithstanding, how many potentially game-changing clean-energy technologies have been developed and invented over the last 30 years, but with poor success rates in converting available technologies to those that have achieved commercial success. Today, one could say that we are sitting in the middle of an innovation logjam [1].

Taking major leading US research universities as an example with several billions of dollars in annual research funding, we see that only one out of every 200 patents filed by university researchers were either licensed or led to a startup. A typical Research and Development (R&D) university is sitting on thousands of such untapped inventions, patents, and technologies, representing a vast portfolio of intellectual property (a similar case could be made for our national labs) [2]. In addition, even in those cases where commercialization was attempted, a majority of such initiatives did not achieve commercial impact or success. This suggests that the Bayh-Dole Act of 1980 for commercialization of technologies developed in US universities [3] has not achieved its full potential. Further, the "Technology-2-Market" (T2M) process that is currently being used has not achieved scale or fueled further innovation. In fact, the success rates of non-software-based (hard science or

D. Divan, S. Sharma, *ENERGY 2040*,
https://doi.org/10.1007/978-3-031-49417-8_7

hardware based) innovations, on a US national basis, have been abysmal. Only one out of every 3600 received Venture Capital (VC) funding, almost 6× worse than for software tech startups [2]—and 90% of them did not reach any sustainable level of commercial operation! This is a very poor "return-on-capital-invested" (ROCI) and it is perhaps not surprising that the already limited VC capital dried up quickly [4].

A deeper look into some of the real numbers helps us better navigate through the complexity of how innovations have transitioned to market in the past, and to understand the implications for today. Does the low success rate mean that the science behind the inventions was flawed and not easy to commercialize, or are there other factors behind the poor results? Does it also mean that there are no useful or viable technologies hiding in the unexploited innovation logjam, and if there are, why are they not being snapped up by existing companies, entrepreneurs, and VCs? Examining the technology stacks, and working with many of the inventors, we have seen that many of these new technologies, *if harvested effectively*, have great potential value, and can contribute significantly to our goals of accelerating science to impact, particularly in the field of energy, and to help us meet our desired future for 2040. Even more interesting is the question of how we can change the way we develop or examine new scientific discoveries and assess which ones can be accelerated to impact. It is in this context that we see these exciting new horizons.

The big question clearly is—what does *if harvested effectively* mean? How do we clear the innovation logjam, and can we change the process to ensure that innovation is quickly and efficiently coupled to the market in a way that major societal energy (and possibly other) challenges are addressed? As we shall see, many (but not all) of these new technologies also feature the steep and sustained learning curves that we have discussed earlier, mainly because they are building on similar core elements. If we can ensure societal value and economic viability of a potential technology (in this case to address a specific energy problem), we can drive scaling based on the economics, while automatically benefiting from a tailwind that helps address the climate issues (again escaping the emotionally charged rhetoric of economics versus climate). This is particularly true in the case of disruptive new ideas where technical viability has not been demonstrated and for which there is no existing market pull.

Developing a process for rapidly commercializing energy innovations, especially those with big potential impact from those disruptive ideas that usually stem from hard science and deep tech, now seems possible and gives us hope for a new era. If done right, it could lead to an ENERGY 2040 scenario that meets all our desired goals. Our ability to rapidly de-risk and commercialize promising new technologies can substantially add to the choices available to us and can accelerate the transition to a new sustainable energy ecosystem.

A parallel evolution has also been happening quietly in the area of commercializing innovation. Along with technology, our understanding and ability to commercialize energy innovations has also evolved substantially over the last 30 years, with many examples of brand-new technologies taking the world by storm. A quick data point stems from the authors' own experience. They have been practicing hands-on entrepreneurs for over three decades in this arena and have had a front-row seat and a lot of learnings—with some successes as well as failures—through this exciting journey. More recently, while working together at Georgia Tech Center for Distributed Energy (GT-CDE) for the last 8 years, they set up a ground-level pilot to develop a process for accelerating deep tech to impact, and to use projects at CDE to validate the lessons, practices, and findings, and in the process, also possibly discover some new insights into the fundamental question of accelerating the commercialization of innovations from deep-domain science and technology. We have learnt much over this period and feel that these learnings have the potential to benefit scientists, technologists, entrepreneurs, directors of research labs, VCs, policy makers, and government.

But before we take a deeper dive into ways to accelerate the T2M process, especially for hard science and deep tech, we need to understand the status quo of the innovation landscape and how it has changed and evolved over the last 50 years or so [4].

Revisiting the Energy Innovation Landscape

Research by its very nature is not linear or transactive. There is no simple equation that tells us that X amount of money will buy Y amount of innovation. Still, at an aggregate level, the amount of money going into the university R&D system produces a substantial output in terms of knowledge and knowhow, but with surprisingly disappointing conversion of the research results into near-term tangible benefits [4–8]. There is no shortage of universities that perform basic R&D. These universities, especially large public and private institutions in the US, are awarded several billions of dollars annually from various federal agencies like DOE (Department of Energy), DOD (Department of Defense), Defense Advanced Research Projects Agency (DARPA), NSF (National Science Foundation), NIST (National Institute of Standards and Technology), NIH (National Institute of Health), ARPA-E (Advanced Research Projects Agency—Energy), etc. Research faculty and students at these institutions generate hundreds of thousands of technical papers along with thousands of new patents and intellectual property every year. Many universities have programs to help generate new ventures based on intellectual property (IP) held by the university. Even with all this, only a handful of new startups see the light of the day. Most will not be successful in developing innovative new products and achieving commercial success, especially when the ideas are disruptive to the status quo. Research and Development (R&D) and basic research pipelines are full of high potential disruptive and transformative innovations, but universities are not able to commercialize the vast majority of such opportunities, and to achieve their full potential.

A similar discussion can be had around national labs in the US. These institutions are supported by the US Department of Energy with a budget of $44 billion for 2021, plus an additional $7.5 billion from the DOE Office of Science to conduct fundamental research. National lab funding exceeds energy-related funding to universities, and while the labs are focused on many specific classified problems of national interest, they also work extensively on commercial issues, such as energy storage, photovoltaics, grid issues, wind energy, and other related matters. Further, national labs do not have the responsibility for education and the issue of "academic independence," as they are, by definition, focused on solving the nation's problems. And they do a tremendous job in generating the science and the knowledge that can guide the nation. Yet, much of what has been said above for university related research and the lack of many disruptive successful startups that are based on translating "deep tech" to market, can be seen to be true for national labs as well.

There are many examples of how R&D investment can produce valuable research that does not go on to create products. For example, R&D investment by the US federal government in non-defense industries over the past 60 years has been about 7 trillion dollars, most of which has gone to universities for conducting research across all disciplines [4]. On average, this has been a total of about 2% of the federal R&D budget. However, if one includes defense R&D over the same period of time, the percentage of the budget spent increases to about 5.5% of the budget outlay, a significant proportion of the entire national budget. Figure 7.1 shows national outlays on research by area, indicating that health and biomedical research represent the biggest category, and energy and climate are a very small portion of the historical research expenditures. Looking broadly, not just at energy, we can raise the same question—*why is our deep tech to market commercializing process so inefficient across the board?*

While this may seem like a huge amount of money, the overall R&D funding as a percentage of the US federal budget has been consistently declining over the past

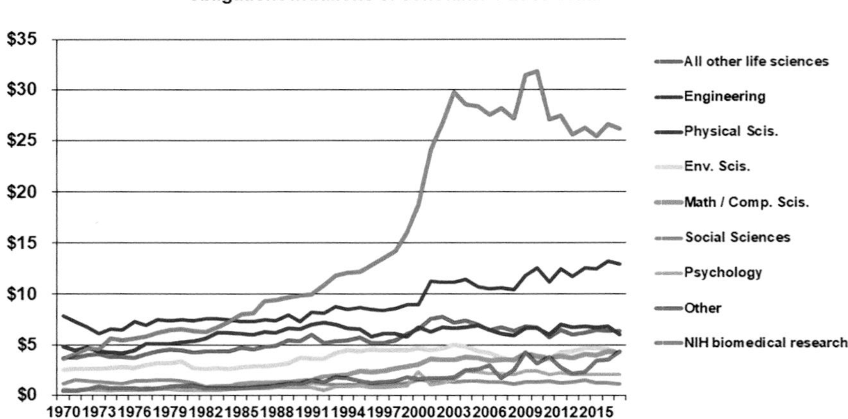

Fig. 7.1 Absolute R&D dollars into most science research has gone up

30 years, especially after the end of the Cold War in the early 1990s. Yet, in the name of academic freedom, universities have traditionally had little pressure to commercialize their inventions. While this may have been acceptable in the 1950–1990 timeframe, the situation, as we will see, is very different now, and *it is important that universities take on a bigger role in converting their innovations to societal value, and to do it more efficiently and in a timelier manner*. Given the urgency of decarbonization, sustainability, and climate change, this is a major societal imperative.

Further, in the era of globalization and the internet, basic knowledge of science and technology has proliferated rapidly, and has become universal. Many countries are now investing more in R&D than the US as a percentage of their GDP, although the absolute amount of dollars may still be lower than for the US [5]. Recent data on national level R&D expenditures (Table 7.1) show China substantially accelerating their efforts and overtaking the US in gross expenditures in R&D (GERD) as compared to what was happening just 10 years back in Fig. 7.2. It is also interesting to note that US scientific effort, most of which is reported in open access journals and publications, helps to inform, and guide the research done in other countries. The result of this has been an increase in innovative products coming out of these countries. This should not be a surprise since countries such as China, in particular,

Table 7.1 Recent R&D spend helps understand changing landscape from 10 years back shown in Fig. 7.2.

Top 10 R&D Spending Countries for 2021							
		2020 Estimated			2021 Forecast		
Top 10	Country	GDP PPP Bil, US$	R&D as % GDP	GERD PPP Bil, US$	GDP PPP Bil, US$	R&D as % GDP	GERD PPP Bil, US$
1	China	29,010.7	1.98%	574.40	31,389.6	1.98%	621.50
2	United States	20,145.1	2.88%	580.20	20,789.7	2.88%	598.70
3	Japan	5174.2	3.50%	181.10	5210.4	3.50%	182.36
4	Germany	4283.5	2.84%	121.65	4480.5	2.84%	127.25
5	India	9991.1	0.86%	85.92	10,870.3	0.86%	93.48
6	South Korea	2002.6	4.35%	87.11	2102.7	4.35%	91.47
7	France	2864.7	2.25%	64.46	2979.3	2.25%	67.03
8	Russia	3927.7	1.50%	58.92	4037.7	1.50%	60.57
9	United Kingdom	2876.7	1.73%	49.77	2983.1	1.73%	51.61
10	Brazil	3199.3	1.16%	37.11	3288.9	1.16%	38.15

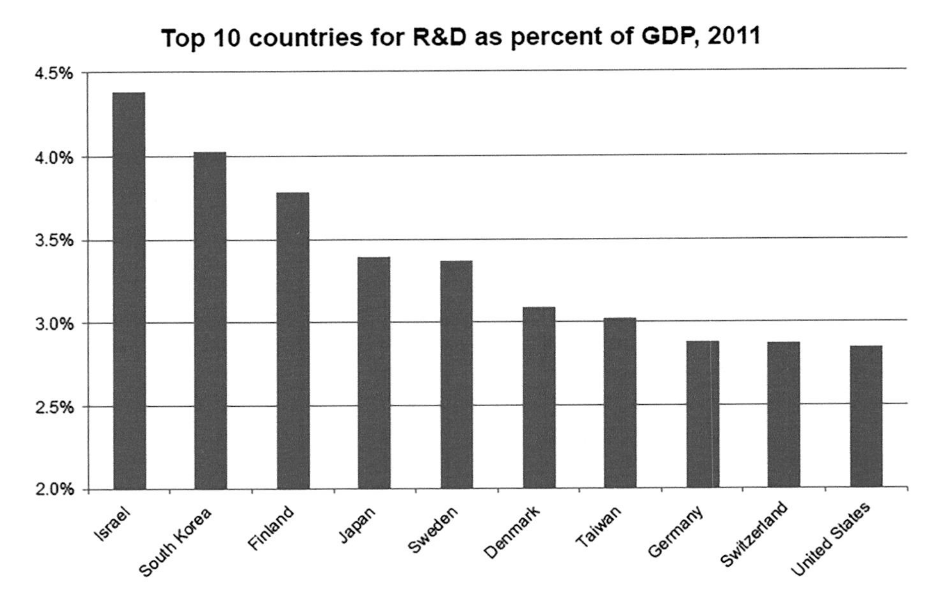

Fig. 7.2 Comparative R&D investments done in 2011 by other countries (https://en.wikipedia.org/wiki/List_of_sovereign_states_by_R&D)

have focused a lot of resources into turning the increased R&D effort into commercial activity and strong GDP growth.

Existing University Metrics Are Misaligned for Innovation Translation

The classic argument for the poor innovation translation output from our university system is that the primary function of faculty has been to teach and do fundamental research. Development is a secondary goal at best, done only if necessary. In fact, faculty members early in their careers are frequently admonished if they do not focus on the primary goals of funding, publications, and teaching—any effort to spend their time on transitioning their technology is often frowned upon. Similarly, at the university level, universities are measured on their ability to draw in students and have a large base of funding and endowments—rarely do these metrics include innovation or impact on solving major societal challenges.

Historically, Energy T2M in University R&D has been a very inefficient process. The numbers tell the story, and "University metrics drive ongoing behavior."

A core principle of organizational psychology is that people will act to improve their performance relative to how they are measured, and as a result—metrics drive behavior. Given that faculty in most universities are measured by research grants received, research publications or instruction-based metrics, with no rewards (brownie points) for innovation or its commercialization, it should not be surprising that most faculty pay little attention to commercializing their innovations [4]. To complicate matters further, the issues of "conflict of interest" (relating to the actual or perceived use of university resources for personal gain) are so draconian at most schools that faculty often prefer to avoid getting into situations that can be difficult to navigate and can put them in personal peril. The schools themselves have had little experience with the translation process, and many venture support initiatives notwithstanding, have struggled with guiding new startups that emerge from their labs—particularly when they were disruptive deep tech to market opportunities. On the other hand, consistent with what organizational psychology tells us, we see our faculty excelling in the areas where they are measured.

As research budgets declined, universities were also faced with the problem of creating more research funding sources and sought alternate ways to turn their research into gold. One solution that came about is the Technology Transfer Office for tech transfer, and an internal Venture Lab or a Studio structure to assist faculty and students with startups.

Startups from Research in US Universities

The Bayh-Dole Act of 1980 triggered a massive increase in technology transfer efforts by universities [3, 9]. Technology Transfer Offices (TTOs) were created to make good-faith efforts to harmonize the process for commercializing university inventions. This was based on the premise that research at any US university was invariably leveraging resources and facilities that were either funded by or subsidized by government resources. As a result, the Bayh-Dole Act required that all university-generated IP be protected and licensed through an institution-level office, the TTO. This process begins with a faculty, researcher, or graduate student disclosing an invention to the TTO, which the TTO then decides to patent and/or license (or not). The TTO then tries to find a partner for commercialization. The partner may either be an established corporation or a startup funded by several possible sources of capital. Although initially most of the activity took the form of license agreements with established firms, there has been an increase in commercialization via startups more recently. However, this commercialization journey has been a slow and inefficient process, with many universities struggling to recover their costs. Over the roughly 40 years since the act was passed, many universities have essentially given up on this as a means of turning their inventions into gold.

The TTO, and thus the university, typically covers the cost of filing for patents, typically in the range of $15,000 to $50,000 per US patent, and significantly more if additional international coverage is required. Public universities usually retain ownership of the patents they pay to register. Over the hundreds or thousands of patents that a TTO may file every year, this can represent significant cost. Given the wide diversity of technologies that a major research university may cover, it is not surprising that TTOs are challenged to market these patents and technology stacks actively and effectively. As a result, the focus of the TTO often is on simply recovering the funds they have expended to get the patents, and not on getting licensing and royalty revenues that can be returned to the inventors and the university to stimulate further research. Even further away is the idea that some of these technologies may be "exponential technologies" that could benefit society and address key challenges of energy, sustainability, and climate change. A few TTOs such as the Wisconsin Alumni Research Foundation (WARF) from the University of Wisconsin—Madison, have done an admirable job in licensing their patent portfolio and returning significant funding back to the university to support further research and innovation, but the vast majority of TTOs struggle with this issue.

Which patents are actually licensed by the university TTO and become part of new emerging products and solutions is largely a matter of chance and the low-probability intersection of an interested entrepreneur, angel investor, and someone experienced in the new emerging technology. *Such a startup team is also generally looking for low-risk quick-to-market opportunities, and not the high-risk deep tech to market disruptive and transformative opportunities that we have been largely discussing in this book.* When the startup team involves the graduate students who worked on the problem, there is technology competency (although not a lot of experience) in the team, and they often struggle with the business model, ramp-up, and scaling of the initiative.

In addition, universities typically do not actively encourage professors (who are often the most competent resource in the technology area) to take an invention and create a new business, especially in the publicly funded universities. Expecting a university to encourage a professor to leave, even temporarily, is clearly against the university's interest. They could lose a competent professor who is able to contribute effectively on the academic and startup side, and even if he or she returns, his or her activities at the university are disrupted during that period. Year over year, even as more patents are granted to universities, only a small percentage are used, and the logjam gets larger. A lot needs to be done to unlock the full potential of university research and to turn it into commercial value.

Data Highlights and Insights

Extensive data on inventions through startups from 1993–2004 was collected by the Association of University Technology Managers (AUTM). Similarly, data on the size and quality of the life science and engineering faculties was collected by the National Research Council (NRC). Additionally, data on venture capital funding has been gathered and presented in the National Venture Capital Association Yearbook (NVCA) as well as the annual NASDAQ composite index. Very rigorous empirical analyses and correlation models have been studied by several researchers over the past 20 years.

Table 7.2 Number of startups vis-à-vis R&D spend in university system [4]

Annual R&D expenditures and startups created

Average annual startups	Average R&D expenditures per startup	Number of institutions
Less than 1	$199,038,199	133
1 to 2	$98,242,994	73
2 to 4	$91,602,939	57
More than 4	$77,050,066	44

Source: Analysis of AUTM data 2003-2009

The results from the analyses are interesting from several perspectives. A snapshot of their findings is shown in Table 7.2.

Further, the data suggests that a large number of invention disclosures have to occur (on average) before one of the patents is used to launch a startup—ranging from about 200 for the full sample data to about 150 in the sample without the University of California system (which accounts for ten universities clubbed into one data set). Regardless, the ratio of startups versus inventions is still very poor, only about half a percent. This means that less than 1 in 100 invention disclosures is used directly in the creation of a product or new venture. Within a university, there can be many factors that affect this pathway from research to product. It can be as simple as a poor relationship of the TTO to the research community or the industry sector that could use the IP, or as complex as nuances of university policy that limits licensing. Even the university's efforts to establish an innovation ecosystem centered on the university can change the results. Some of these are foundational to fixing the innovation logjam.

Industrial Corporations No Longer Do Long-Term R&D

The post-Cold War era of the 1990s and 2000s also changed the American corporate strategy to one that was closely tied to creating immediate shareholder value. Companies lost sight of the long-term sustainable competitive advantage that comes from R&D. This shift led them to have a very short-term profitability and growth mindset as desired by Wall Street. This behavior, in the absence of the large federal grant dollars disbursed during the pre-Cold War era, quickly resulted in the virtual death of any disruptive or game-changing industrial R&D in mature industry sectors such as the energy industry [4].

Within leading corporations, the people, data, capability, and market access—all exist. After all they are very successful in growing their existing product lines and businesses, typically doing very incremental innovation reflected in adding new features or functions to gain incremental market share and improve profitability! What is generally lacking is the proverbial startup or entrepreneurial culture, and an appetite for transformative innovation that could have societal impact and scale but could also be disruptive to their existing business. Once again, their internal metrics are not aligned with rapid deep-tech innovation and disruption. Innovation risk is not rewarded unless it is part of the "planned budget." In fact, the typical large company has no realistic and established way of adopting innovations that disrupt their primary business models [4].

If industry has not done long-term research for a long time, how do they instantly catch up to a new fast-changing reality that is disrupting all their investments, beliefs, and the markets in which they have traditionally been leaders? It is no surprise then, that many of these entrenched market leaders are struggling to adapt to the furious change that we have today, especially in the energy industry, with many trying to reinvent themselves as the megatrends become clear, while even more are still sticking to their old paradigm, making perfunctory incremental changes, and hoping that this day shall also pass.

It would be naïve to believe that establishing these longer-term research mechanisms alone will make innovation and its translation to market magically happen. Encouraging and sponsoring innovation is hard, constant work. What makes things worse is when the industry market leaders work in close conjunction with regulated organizations, such as utilities, who are also seeking to slow down the pace of change. It is striking that even with all these incumbents pushing back on disruptive change, the change is occurring fast and at a pace that seems impossible to stop.

To succeed in bringing innovation to the corporate world, one must work with the culture that exists. We do that not because the culture is broken, but because it is not broken, and it is what keeps the corporation working. Simply demanding that people learn more and try to think innovatively is not enough. There must be fertile ground prepared by a leadership team that really considers innovation and company reinvention a key part of their strategy and builds a community of innovation within the company and supports pathways by which innovations can be adopted and scaled.

This exposure, both to the problems and the potential means of solving them, is critical and is typically absent in large corporations, especially the established leaders in a highly regulated energy industry. To paraphrase an ancient saying, vision without knowledge of the problem is a dream, action without vision is a nightmare [4]. Incumbents who control the market resist disruption because corporations do not want any business risk. That is the highest risk of all and simply guarantees the nightmare scenario, especially when things around you are moving fast.

Most corporations, like every innovation ecosystem, are in constant search for "new" ideas, but only those that are not disruptive to their business and investments they have already made. While they may be willing to take on some level of business risk, they are certainly not willing to take on "unproven" and "poorly performing" technologies and business models—typically the characteristics of disruptive technologies in their early stages. It is not surprising that "intrapreneurs" and skunkworks seldom succeed in a large corporation.

Another classic example that demonstrates industry commitment, and their approach, to searching for new ideas is a proliferation of Corporate Innovation Hubs in Tech Parks that are being located near prominent research universities [10], or an increasing number of university-industry consortia that work on advancing relevant technologies. Industry either closely guides the research so that the results are not disruptive to their business or has no idea what to do with any disruptive ideas that may be developed and presented to the industry partners. It is clear that large corporations, without externally subsidized or incentivized R&D, are poorly motivated to disrupt their own business, a trend we have seen repeated ad nauseum in the ongoing energy transition.

Venture Capital Does Not Like Innovation Risk Either

A focus on achieving the promised Return on Capital Investment (ROCI) for their limited partners (LPs) over the period of the VC fund (typically 7–10 years) drives the overall culture and behavior of venture capital [4]. With few exceptions, most VCs avoid innovation-risk, risk that feels open-ended with multiple factors that can deny success. They like to be at the "leading edge" and not the "bleeding edge," with many "fast followers" who typically jump in once the new opportunity is validated and innovation risk is eliminated. Coming with a fresh pool of capital, these firms can speed up to overtake the wounded and often tired firm that was at the bleeding edge. We often see "herd" behavior where VCs will move in groups to share the risk in many different approaches to tackle a specific opportunity. This pattern is constantly repeated in Silicon Valley, whether it is the latest in AI or solid-state batteries. We need to realize that the VC model is a financial return model, and that innovation is the engine on which it rides, not the other way around. Their motto is "sustainable competitive financial advantage" through differentiation, not necessarily technology differentiation, or through disruption.

Along with undesirable levels of risk that come with some of these nascent technologies, regulated markets, such as electric utilities, pose additional risk in terms of painfully slow adoption rates—something which makes the returns in this sector very challenging. Seeing the massive opportunity that the energy transition represents, VCs keep jumping back into the energy space, hoping that something has changed since the last time around. Unfortunately, in the grid space they are still seeing scant success due to inordinate delays. Such delays burn up excessive amounts of capital, wiping out returns for the VCs and the founders of the startups, often leaving companies vulnerable to a hostile takeover even as they are on the brink of success.

To deal with all the unknowns and the poor success rates, VCs focus on identifying the few companies from their portfolio that are likely to succeed and become "unicorns," with the magical $1 billion in valuation and hopefully outsized returns. It is very likely that the remaining companies, even if they have real products and sales, will either be sold for modest or no returns, or will be shut down—in most cases with the founders and employees getting little for their efforts. The VC model looks at averaging the returns across their portfolio, mainly depending on the unicorns to get the outsized returns that they have promised their LPs. This does not mean that the companies that did not become unicorns had poor technology or business models—maybe they were focused on the energy sector where the customers did not move fast enough. There has to be a better way to de-risk an investment, particularly in the area of deep-tech disruptive opportunities that are the foundation for the ongoing energy transition.

Even with the very high failure rates and poor performance that we see in the VC model, these models have survived until now, and actually work well enough for incremental technological advances. It is the big bets, such as with Tesla, that have provided great returns and have moved us closer to our societal goals. New VC models are emerging to tackle the areas of higher risk to tackle societal challenges, but these are few at this time. It is important to put a spotlight on the inefficiencies of capital deployment and value creation—across the complete investment cycle, from deal sourcing to exit—and to see if corrective measures are needed. In good times the investors generally receive the promised returns from Venture Capital firms, but what was left on the table and did not "succeed" is frequently not questioned nor properly analyzed. We feel that this is an unsustainable trajectory to commercialize innovation and better models need to be explored.

Venture Capital Is the Oxygen for Innovation

Venture Capital has played a critical role in the US to keep the innovation industry going and has been instrumental in making the US the leader in new digital and related technologies, areas that are only lightly regulated and can move at the speed of innovation and commerce (not at the speed of regulation). From its original objective of funding technology companies that create new value, Venture Capital (VC Funding) now approaches the business of startups as an "investment vehicle." This is especially visible when one looks into investments in the rapidly growing tech startup community. Thousands of startups receive some level of investment every year. It is these startups that are the fountainhead of innovation.

On the surface, the macro picture looks very promising (with as much as $80B invested in peak years) but is primarily focused on digital and IT products and software. The investment into energy hardware is much lower, as can be seen in Fig. 7.3, mainly because of higher risk and longer adoption times. When we look under the surface, the startups also have to deal with a lack of process-driven investment mechanisms and strategies, which are supposed to, at least in principle, help each and every one of them realize the full potential that they offer. The ideal is to then have this support available and providing value for the large number of unique and highly differentiated startups that are in the VC ecosystem. The existing ecosystem system is too limited in its geographic reach and is largely centered on the persona of the venture fund managers, a model that generally results in fewer and larger deals. VC capabilities need to be amplified to allow more deal flow and to achieve a higher chance of success [12]. Incubators and accelerators are trying to fill that gap.

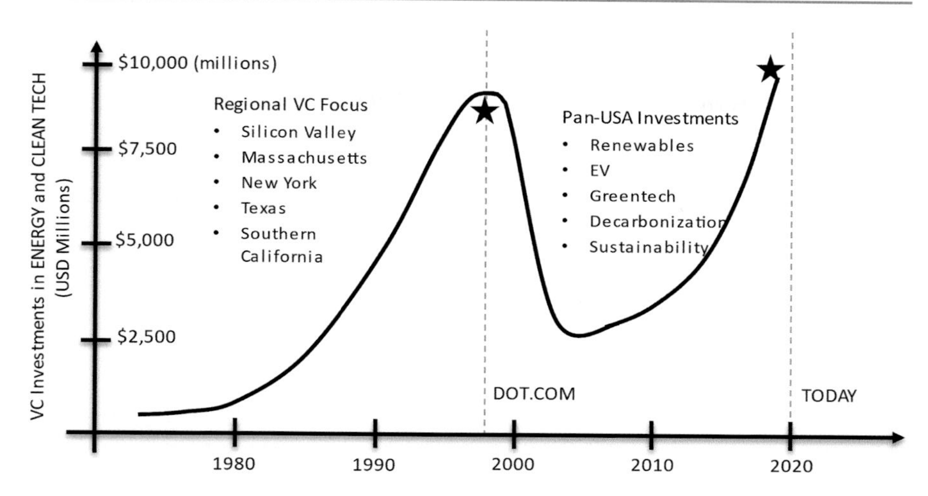

Fig. 7.3 VC Investments since 1980 in energy. The peaks in graph show Dot.com boom, and now [11]

Incubators and Accelerators—The Entrepreneurial Ecosystem

Currently, there are about 9000 incubators and accelerators around the world [13]. Almost 70% of these are in the US. And this number changes by the week, if not by the day. A variety of business models can be found among these incubators and accelerators. Many are focused on specific industry sectors. A detailed contemporary study can be found in literature [4]. There are very few that focus on the energy industry, but that may change going forward as the rapidly growing new opportunities become clear. More recently, with the focus and federal funding coming into clean energy, sustainability, water, environment, and climate—a new crop of dedicated energy incubators and accelerators are popping up.

However, the harder question to answer is whether these innovation hubs reflect a repeatable commercialization process for disruptive innovations, particularly in energy, to help these budding startups achieve scale, speed, and impact. Our key findings from our own entrepreneurial experience as well as from a study of almost 4000 such entities reported in the literature are: [4]

- Understanding "patterns" of successful startups is not enough; we have to look at the patterns of the failures as well. They are equally essential to develop a repeatable and robust process for rapid commercialization.
- De-risking is not a step function but is needed continually along the way and is a multidimensional iterative 24/7 process.
- What kind of physical and intellectual infrastructure is needed to create an effective innovation ecosystem? Can it be done on a university campus, and/or in an R&D lab? Who are the leading role models in this endeavor?

- National labs do participate in core research and have some success in coupling their research to industry through Cooperative Research and Development Agreements (CRADA). However, when it comes to coupling disruptive solutions to market, their success rate, like their university counterparts, is also poor.
- The traditional approach of "tinkering" (incremental innovation and differentiation for competitive positioning), once major science, technology, and innovation risks have been removed (as discussed in Chap. 2), is still the common practice. Sometimes, learnings from the market are positioned as "pivots," often without seeking a holistic and deep understanding of the real problem or the customer needs, which if done upfront, could result in faster commercialization.

However, most of the incubators and accelerators have little to offer on the technology side. They tend to assume that the startup team has the technology skills and are focused on teaching them business and entrepreneurial skills, including customer discovery, market and competitive assessment, product positioning, business model development, and preparing the pitch for investors and venture capitalists. Most of them assume that the market knows what it wants—which is possibly true in mature industries where the startup is trying to gain competitive advantage, but frequently does not hold when new disruptive ideas based on unfamiliar science and deep tech are being proposed.

Conventional entrepreneurial programs in business school, or in most incubators/accelerators, mostly focus on the mature industry segments where there is a broad understanding of the market and competition, and often assume that the deep-tech disruptive startups will have to deal with the same issues, and so train them accordingly. As we have discussed earlier, major gaps exist in our understanding of the deep tech to market roadmap. Some of these are discussed next.

What Is Deep Tech and Why Is It Different?

Deep tech is about potentially disruptive or transformational innovations or inventions that are a by-product of a large body of cutting-edge R&D work that has overcome scientific and technological barriers but is waiting to leverage appropriate engineering and scaling to impact an established industry sector, or to create a brand-new industry sector. By its very nature, the terms deep tech and disruption suggest the following:

- An absence of established market-understanding or pull, and a need for new insights that can accelerate adoption.
- Lack of a broad pool of potential employees who understand the new concepts and technology.
- An abundance of experts in the existing technology and market sector who do not understand the new disruptive concepts, and will either push back on proposed solutions or misunderstand its potential use cases and impact.

- Poor knowledge about factors that will influence the manufacturability, reliability, and scalability of the proposed solutions.
- Poor understanding of cost, trade-offs, and timeline for product development and release.
- Uncertainty about key factors that will limit its performance.
- Lack of existing standards, policies, and certification that will impact scaling and timeline.
- Need to grow a workforce trained in the new technology.

The list goes on and the factors are all very important. This also indicates why the risk in deep-tech startups is so high. It also explains why the almost chaotic process of "tinkering" by hundreds of parallelly operating teams has been the only "strategy" that has worked, and why such industry or market transformations have been very expensive and have taken 75+ years in the historical past.

What we are seeking is a dramatic acceleration of the process by which deep-tech transitions to market, to scale, and to impact. With the universalization and global explosion of knowledge, such acceleration can provide strong competitive advantage. It can also shorten the timeline in terms of tackling major societal problems, such as sustainability and climate change. What is clear is that deep tech requires us to reexamine the normal go-to-market paradigm.

What we are seeking is a dramatic acceleration of the process by which deep-tech transitions to market, to scale, and to impact.

Traditionally, in a startup, we keep the core technology team small and focused so that burn-rate is low in the early stages, bringing in experts from outside on an as-needed basis to address non-technology issues, such as operations, manufacturing, finance, marketing, and business strategy. High-impact transformative deep tech may have been developed by big teams in well-funded research labs, but in the startup setting, now has to operate with a small team that may be relatively inexperienced in the deep domain and other areas. If we fill the gap with outside experts on a consulting or part-time basis, they are not experienced in the new technology. Further, the core technology team generally has poor understanding of the translation process or industry needs. Taken together this is a very challenging situation. This process leads to many misfires, dead ends, and pivots—resulting in lost time and money, loss of equity for the founders and early investors, and lost competitive advantage. Unfortunately, this is often the norm in many deep-tech startups and is one of the reasons for the abysmal success rates, and the hesitancy of VCs and investors to invest in deep tech. We need to systematically understand the stages of this roadmap, which, at a high level, looks similar to the traditional roadmap, but may require a fundamentally different de-risking process when it comes to execution.

We generally begin the exercise of transitioning a technology to market once we have seen that it offers unique value streams and has the attributes that could meet key market needs. In many cases, once the technology is introduced into the market, the market itself tells us what the unmet needs are (remember the improvements on the Savery steam engine, each of which addressed an identified need and shortcoming of the original solution).

In the case of deep tech, the market is not asking for anything, and researchers are typically guided by their science interests and a fuzzy and narrow basis for how and what value it may be able to provide. In many cases, these are large teams working on multi-year programs to crack the hard science problems. As the scientific breakthroughs are identified, maybe a microscopic scale demonstration unit is built—and victory is declared. Inventions are disclosed and patents filed by the TTO, and students graduate and move on. With many such teams working on different methods to address the basic need, a possible entrepreneur interested in commercialization has a high hurdle to climb (remember that the high innovation risk and potential disruptive nature makes these a poor fit with larger market leaders). The entrepreneurs have to understand multiple deep-tech concepts from different teams, compare them, ascertain if any of them have hidden flaws that have not been disclosed, identify key barriers to commercial success and scaling, and do all this without any funding or access to the facilities and team that developed the concepts in the first place. It should not be surprising that this process has a very low success rate.

Current Processes for Science to Tech Transfer

As we look at the innovation logjam at research universities and laboratories in the US, it is clear that there is a substantial amount of innovation and knowledge generation that is taking place, but that it is not well coupled to actual solutions that can address some of the pressing problems that we, at a societal level, are facing today [12] [14],.

Organizations such as NSF, ARPA-E, DOE, DOT, NIST, DOD and others are recognizing this core issue and are trying to pay attention through programs such as I-Corps, Technology, and Innovation Partnership, Small Business Innovation Research (SBIR), Small Business Technology Transfer (STTR) to name a few, all aimed at improving the Technology Readiness Level (TRL) of emerging technologies. I-Corps has allowed scientists to gain experience in the process of creating startups by working with mentors in areas of customer discovery and testing their solutions with potential partners and customers. Over 1000 startups and $760 M in funding raised attest to the need and to some momentum. Yet, the process, like most of the university venture programs, is guided by traditional methodology including customer discovery, fail-fast and pivots to help align with, and meet market needs. This is very important for incremental technology where the market knows what it needs, but can be detrimental, giving wrong answers, if applied to deep-domain research that the market is unfamiliar with, and where the new technologies are not currently commercially viable or competitive (remember PV solar in 2000!). The issues of steep and sustained learning curves as being a key driver are often not considered.

Incremental advances in technology (such as improved functionality or use of new materials to improve existing products) are well handled by current processes, including university-industry consortia, NSF Engineering Research Centers (NSF-ERC), national lab Cooperative Research and Development Agreements (CRADA), SBIRs, and STTRs and work being done by venture labs, TTOs, incubators, and accelerators. The DOE Loan Programs Office, as well as many of the new Biden government initiatives, such as the Bipartisan Infrastructure Law, Science Act, and CHIPS Act are aimed at increasing US manufacturing of existing technologies, and do not necessarily apply to high-risk new technologies that are sitting in our advanced science and disruptive technology research labs. Yet, we have seen that many of the old twentieth-century technologies cannot incrementally be extended to realize twenty-first-century objectives. New solutions are needed, and they are needed now. *What we need is to be well ahead of the market, anticipating their needs, and charting what the path to scale is.*

Summary of Existing Gaps Across Various Commercialization Ecosystems

We have identified and discussed what we feel is a major gap in the "science to market to scale and impact" process that exists today, especially for deep-tech science and disruptive technologies.

From a holistic process standpoint, it is worth reiterating and summarizing the following:

- Post-Cold War, the USA saw the old industrial-military complex gradually veer away from long-term R&D. In the process, the coupling of fundamental research with industry was weakened or broken.
- In parallel, the Fortune 500 culture also skewed their metrics toward meeting quarterly Wall Street expectations of near-term shareholder value. This resulted in a focus on small, incremental innovations to protect their own market share, and the elimination of long-term R&D.
- Entrepreneurs interested in deep tech were also handicapped, especially when they needed deep-domain knowledge access. It was an even bigger problem for startups that were not based on coding of an App done by "a few people in a garage," but were focused on special new materials or hardware, and required special expensive equipment to build or test the products!
- The traditional VC community often viewed disruptive energy startups as highly risky "science projects," with huge challenges in reaching scale on a reasonable timeline and stayed away from investing in them.
- Most VC firms would not really take on innovation risk that they did not know how to manage. There is also an acute shortage of seasoned operator-investors.

- Over the last 30 years, although the incubator-accelerator entrepreneurial ecosystem was evolving [4], it also did not fully know the "process to commercialize deep tech." Extending models for traditional entrepreneurial startups to deep tech resulted in high failure rates, and reduced the interest in such opportunities.
- At the same time, the place where all the deep tech was located and being generated and where they had the competencies and the equipment, i.e., the university R&D centers and the national labs, remained focused on fundamental science, counting on someone else to do the translation. The government kept increasing funding and trying to increase translation efficacy—but results were poor, because fundamental issues with the model and metrics were not addressed.

This is where we find ourselves today—rich in science, but with poor efficacy in the translation process.

What is disconcerting is that, because the knowledge is universal (as it should be), other countries are learning from us, and are setting up more effective translation processes. Even more concerning is the question of how many potentially valuable technologies that could address huge societal challenges are lying undiscovered in the innovation logjam, and what we can do to improve the processes.

We have to find a better way to translate science to impact! We will discuss one specific example of how we have tried to accelerate the transition from science to market—we call it "Innovation-by-Design."

Accelerating Deep-Tech to Market: Understanding the Requirements

Innovation-By-Design—An Example

We have been exploring an alternate approach on a pilot basis for the last 8 years at the Georgia Tech Center for Distributed Energy (GT-CDE) where we have been looking at the viability of a distributed and decentralized grid as a key enabler of a future decarbonized energy system. We have seen that such a development represents a paradigm shift and could be highly disruptive for the regulated electric utilities and their vendors. But as we have seen, such capability is critical for the upcoming energy transition. However, when we started our journey, little was known about what such a grid could look like, what devices would be needed, and what the individual and collective control strategies could be, how practical, scalable, and resilient they would be, and what the roadmap for the transition from today's system to a future system would entail. As we have seen, nobody in the electricity industry was looking for such a change, and virtually nobody from the operating companies understood the issues such a transformation would have to address—particularly at a detailed implementation level. We were speaking a different language from the "experts," and there were no viable and accepted business models. Finally, such a decentralized grid could collide with the existing $3 trillion that has been invested in the existing US centralized grid, resulting in a high level of pushback and reluctance to change—all factors we have seen and examined in previous chapters of this book.

We understood early on that *we had to reduce the innovation and technology risk at CDE itself*, a lab where we had the technical capability and equipment that few startups could afford, and where we also understood the market needs and gaps—both, at the current point in time, and also when projected into the future. We understood that more important than the final end-goal, perhaps, was a viable pathway to get there, and the recognition of an urgency to develop solutions that were flexible and could adapt as other adjacent or competing fast-moving technologies evolved at the same time, but with no coordination with each other. *We understood that we had to, internally within CDE, take the next step to translate the core science principles into a "value" based solution that people could understand and assess the viability of the proposed solutions at a business level, without their having to become experts on the technology itself.* It was also clear to us that not all projects would translate into high-value opportunities, some were important from a knowledge perspective—and that was fine given our academic mission. It was perhaps as important to know what should not be commercialized today, as it was to identify and accelerate those opportunities that were poised to scale.

> *We had to reduce the innovation and technology risk at the research lab itself, where we had the technical capability and equipment that few startups could afford, and where we also understood current and future market needs and gaps.*

We found that *if* this "translation" step was done sequentially after the basic scientific principles were proven (as was typically the case), we had poor alignment with market needs, and as a result, poor outcomes (Fig. 7.5). Rather, we developed a process of "*holistic and continuous storyboarding*," where we would ask both fundamental and broad questions, including inputs from science, technology, and market, to evolve a tech-agnostic vision of what such new capability could do in the market, why it was needed and important, particularly at scale, and if it was worthy of serious attention. From that we would evolve what the deep-tech requirements were, and what adjacent issues and technologies could become showstoppers (such as existing policies) and had to be *simultaneously* addressed. If the proposed solution that emerged from this process was incremental, we could see how it would position against other solutions and follow a more traditional T2M path, possibly licensing the technology to a big company. If it was new and disruptive, we would try to understand what new capability would be unlocked (if the technology did what we thought it would), why it was important at scale, and what key functional requirements were needed at the technology level, and finally if the technology could ride a steep and sustained learning curve to get to economic breakeven along an accelerated timeline (Fig. 7.4).

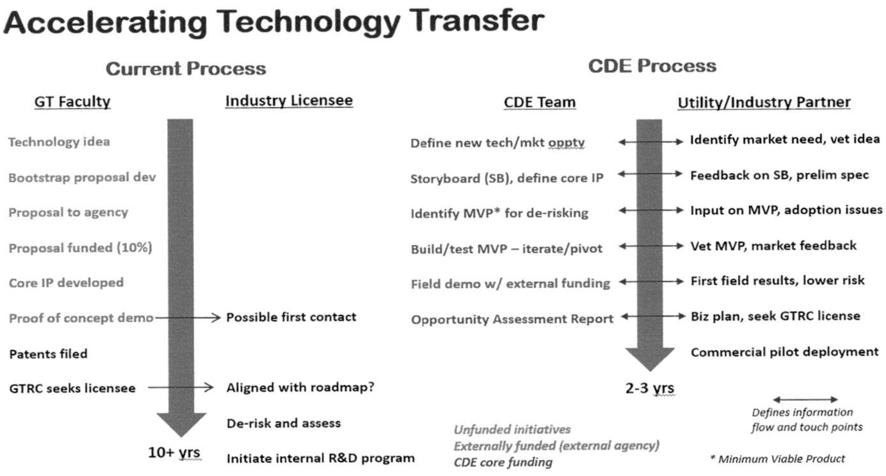

Accelerating Technology Transfer
Current Process
GT Faculty
Industry Licensee
Technology idea
Bootstrap proposal dev
Proposal to agency
Proposal funded (10%)
Core IP developed
Proof of concept demo
Possible first contact
Patents filed
GTRC seeks licensee
Aligned with roadmap?
De-risk and assess
10+ yrs
Initiate internal R&D program
CDE Process
CDE Team
Utility/Industry Partner
Define new tech/mkt oppty
Identify market need, vet idea
Storyboard (SB), define core IP
Feedback on SB, prelim spec
Identify MVP* for de-risking
Input on MVP, adoption issues
Build/test MVP – iterate/pivot
Vet MVP, market feedback
Field demo w/ external funding
First field results, lower risk
Opportunity Assessment Report
Biz plan, seek GTRC license
Commercial pilot deployment
2-3 yrs
Defines information flow and touch points
Unfunded initiatives
Externally funded (external agency)
CDE core funding
* Minimum Viable Product

Fig. 7.4 Comparison of existing tech transfer (T2M) process vs. Innovation-by-Design

CDE – New Model for Innovation & Acceleration
Accelerating Innovation on a Public University Campus
Innovation by Design
Fundamental Breakthroughs
Exponential Technologies
Market/Tech Knowledge
Industry/Market Input
Storyboard
- Align T2M
- Mkt Potential
• TAM
• Time
• Cost
• Value
• Viability
• De-risk
• Incubate
• Accelerate
• Sustain
• Leverage
CDE
- Shared Brain Trust
- Shared Maker Space
- Leveraged $$
• Technology
• Market
• Regulatory
• Business
Near Term Incremental
Near Term Disruptive
Mid Term Transformative
Academic Research & Edu.
• Indemnify
• Onboard
• Mentor
'CDE Labs' Commercialize
• Startups
• Product 2 Mkt
• Evergreen Fund
Industry Solutions
Augment
Disrupt
Transform
Center for Distributed Energy – Innovation Process (2-3 years)

Fig. 7.5 Innovation-by-Design process can drive significant acceleration of the commercialization process across-the-board, even in a publicly funded university

A visual flow chart along with key functions being performed along its translation journey of bringing deep tech to market, as developed at CDE, is shown in Fig. 7.5. This process resulted in a "target specification," and a first glimpse of what a possible "minimum viable prototype" (MVP) could be that we could use to showcase the new functionality and value to a reluctant end-user and/or customer. This ensured that even in the early days of basic research, the solution (provided it eventually worked), was already aligned with potential market needs and impact at scale. During this time, we worked closely with industry (end-users, not R&D folks from

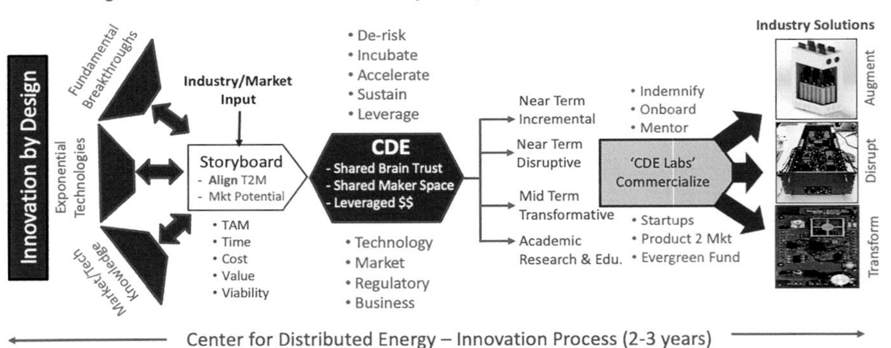

companies that were leaders in the current technology) to iterate quickly on value and features/functionality.

The goal was to demonstrate core functionality, pathway to manufacturing at scale, the value stack that was enabled, and to then build a reasonable scale MVP model, with a final iteration of the design for a first "field-test" for validating its core new functionality. This also allowed us to answer some of the key questions that potential entrepreneurs, partners, or investors would ask, ensuring that the highest innovation-risk issues were addressed, and that tech transfer could be easily accomplished. When we encountered a major hurdle, which we often did, we went back to the storyboarding process. We always considered that finding answers to questions that others in industry had posed were not as illuminating as uncovering new questions as we probed through such a complex multilayered problem definition process. *It is interesting to note that because we were asking questions that were very different from what others in industry were exploring because of incrementalism and a desire to not disrupt their own business, we almost always wound up with key insights and solutions that looked very different.*

We heavily leveraged our work with research grants from DOE and ARPA-E, internal grants through the Georgia Research Alliance, as well as from select industry partners where we had connectivity at the strategic and executive levels. Right from the start of a potentially deep-tech disruptive project, until it was demonstrated in the field with a business model wrapped around it, the project was under CDE control, where we could make quick decisions and provide sufficient funding to move it along to a point where the prototype could be demonstrated, and the value proposition and viability in the market were validated. Discussions with possible end-users and/or potential partners and customers further confirmed the value proposition and scaling pathways.

It should be noted that failure was very much part of this process, sometimes delaying a successful outcome by months—but that was okay and an important part of how we learnt. Unlike in an investor-funded endeavor, we had the luxury of being wrong without the pressure of explaining why we had not achieved the desired outcome, often accompanied by a loss in valuation and equity. Such an integrated approach provided a significant level of both technology and business model de-risking, enhancing the chance of success and the value of the new technology, even for disruptive scenarios.

While this is certainly not the only way to de-risk deep-tech startups, we have seen several startups come out of our approach (with many more in the pipeline) and have learnt many important lessons along the way. The importance of nurturing the translation process within the deep-tech center cannot be overstated, while at the same time, maintaining a "forward-looking" view of what possible market scaling could entail, and what cost/performance metrics would be needed for success at scale. It was important to also consider policy questions, mainly because alignment could substantially accelerate adoption and scaling, and because policy barriers to scaling could suggest alternate ways of thinking about the issue or could slow the process down and needed to be identified early on.

What was striking about this new "Innovation-by-Design" process was how different the very questions, and thus the answers, were from what traditional R&D labs and researchers were considering. At the same time, we did not lose the academic integrity and rigor, as evidenced by quality of publications and PhD thesis work. *Rather, we broadened the connected universe of questions, and framed the questions so that the value was discernible and became an integral part of the discussion.*

We also created funding strategies that allowed us freedom to develop and mature the basic inventions within CDE with no strings attached, filing patents to protect core concepts generated through the GT TTO (Georgia Tech Research Corporation—GTRC), leveraging outside funding through aligned projects, creating sustainable funding strategies for the academic center through licensing of IP generated, and guiding the TTO in terms of tech transfer partners and/or startups. This also allowed us unprecedented speed of execution helping to further shorten the timeline of science to impact.

Challenges remain, including—new mechanisms to provide funding for this translation process; efficient management of Conflict of Interest (CoI) issues; the ability to field-test solutions developed by an academic center, especially around the question of indemnification; and the need for a TTO licensing process that is not based on cost recovery but on value created. While CDE itself works in the specific field of solutions for a future distributed and decentralized grid, the underlying concepts can be applied to a much broader range of technologies and areas. Coauthor Suresh Sharma has been taking this message outside CDE to a variety of other Georgia Tech research centers, including in the areas of direct air capture, green hydrogen, nanomaterials for energy, electrification of industry, energy storage, and plastics from non-fossil fuels—with very good response and acceleration.

Creating a Process for Innovation-By-Design

The preceding discussion suggests that the ability to rapidly and successfully commercialize deep tech presents a truly game-changing opportunity to accelerate the translation of deep tech to scale and impact. Unfortunately, many of the R&D teams ("inventors and innovators") do not typically have all the competencies and resources that commercialization requires. Relying on traditional T2M strategies and metrics that are more suited for incremental innovation can lead to suboptimal results when applied blindly to disruptive deep-tech opportunities.

The process for deep-tech translation assumes that much of the conventional wisdom may not fully apply in the new paradigm and focuses on asking more fundamental questions. A "T Model," that looks deep and broad at the same time, is needed to assess technology requirements in the deep-domain areas that are the strength of the team, but also in the broad adjacent areas that can become showstoppers in a successful translation. Commercial opportunity assessment should not be "bottom-up," where we have a solution looking for a problem. Rather, it should be a simultaneous top-down and bottom-up process that helps to define those "magical" wishful opportunities that conventional solutions cannot address, that are enabled by the new deep-tech capability, and where the potential value and impact would be high. It is also important that the top-down vision drive the technology assessment and gap identification process. It is critical to have the ability to look dispassionately at our own solutions, and to decide if the effort to commercialize at this point in time would be the best use of resources.

We can crystallize Innovation-by-Design as a strategy for accelerating deep tech to market initiatives through technology and business de-risking. It seems to be best suited for use in academic and national lab research centers (not industry labs where near-term non-disruptive product development is generally the primary goal), based on the following elements:

- Strong multidisciplinary competency, including deep-domain knowledge in one or more areas, that looks holistically at complex problems that can have impact at societal scale, defining objectives and framing of the problem (the "North Star") using a technology-agnostic value-based process.
- A "T-Model" for competency to ensure that all deep-domain as well as adjacent areas that can impact viability of solutions (both early on and at scale) are included in the discussion as early as possible—factors such as current technology and market status, as well as gaps and velocity of change in the adjacent areas.
- Strong contact with industry and a deep understanding of the existing market to allow development of scenarios for how these new disruptive technologies with steep learning curves could enter and scale—this could be through existing players or through startups.
- An organizational structure that allows rapid decision-making and action, is supported by sustained flexible non-dilutive funding mechanisms that preserve equity, provides resources to help accelerate decision-making and MVP build/assessment, and encourages a culture that embraces real facts that often disprove our own theories and allows pivots.
- Knowledge pursuit and research are still the primary mission for the center, but the framing of the questions is now done in a way that both the "scientific" and the "innovation for deep tech to market" goals are achieved at the same time, rather than sequentially.
- Rapid protection of fundamental new concepts, ideas, and innovation using "provisional patent disclosures" followed by open publication furthers both knowledge and academic goals, while still preserving competitive advantage for the technology if it goes to market. The commercializing entity can obtain further strategic patents to strengthen the foundational patents.

Modifying the innovation process as discussed above fosters research that meets and exceeds academic goals. It also forces us to embrace the tough open-ended questions that should really be the domain of academic research, but at the same time, acknowledges that accelerating deep tech to market for the right type of technologies (also at the right time) can have high societal value, and is one of the reasons that research funding is being provided.

The minimum viable prototype (MVP), detailed product requirement document (PRD), and a preliminary go-to-market model demonstrate that an implementable solution is possible that solves a tough previously unsolved (often unidentified) problem, but which also overcomes key innovation risk questions, and simultaneously addresses and de-risks issues related to technology, production, cost, scaling, and policy. Scaling and growth and its possible dependence on steep and sustained learning curve are addressed head-on. Investor and commercializing/manufacturing partner inputs in the early stages can help ensure that key relevant questions that are precursors of success are addressed early on, reducing risk for investors to "normal" levels, even for high-potential disruptive opportunities. Not only is such an approach beneficial for the early investors, but also for the founders and technology innovators, as it preserves equity in the startup/technology, reducing the level of dilution needed to scale the company.

We feel that there is potential for national labs and other universities to bring several other disruptive technologies to market by suitably adapting the "Innovation-by-Design" model discussed here. In some cases, a regional consortium approach may work better, but the underlying principles to commercialize are likely to be similar. The authors are currently working to develop functional and operating models to scale this CDE process across university and national lab infrastructure.

While we were formulating these processes at CDE focused around the area of distributed and decentralized energy solutions, it was clear that the lessons learnt were relevant across a broader cross section of technologies, and also raised fundamental questions about the role of universities and national labs in a fast-evolving global ecosystem of knowledge, technology, and industry. Those deep questions are for another time and place.

Climate Change Provides a Tailwind

From the advent of the industrial revolution, until today, approximately 2.4 million megatons (2400 gigatons) of CO_2 have been released into the atmosphere, of which 950 gigatons were absorbed through the natural CO_2 cycle. That still leaves an *additional* 1.45 trillion tons of anthropogenic CO_2 in the atmosphere, roughly equivalent to 700 times the total weight of all the fish in our oceans, and all a result of human activities [15]. Thus, the modern new wealth of nations and their economic prosperity comes at a price. The warming of the planet, currently spread across all continents, major oceans, and twenty-four time zones, is the by-product of heavy industrialization primarily driven by fossil fuels. This dynamic is both the cause of our prosperity and good fortune, but also could be the means of our probable destruction. The damage is in large part the work of Adam Smith's invisible hand, guided by the belief that money buys the future. But we know that nature does not take checks. Who then pays the piper—does infinite consumption survive climate change, or does a changed climate put an end to this uncontrolled growth? [15].

But as we have seen, the tragedy of the horizon has prevented us from collectively acting on long-term climate issues. We have also seen that "exponential technologies" with steep and sustained learning curves offer a once-in-a-lifetime opportunity to achieve a future energy ecosystem that is both lower cost and good for the planet. To achieve the impact in a timely manner, that is over the next 15–20 years, we will need a broad palette of technologies, many of which are still in the laboratories. Looking historically, we see no precedent for such rapid translation from science to impact in the past, but with the onset of the twenty-first century, we do begin to see many examples, all based on exponential technologies (fortunately, well timed for our need to act on climate issues).

If we can quickly develop and demonstrate viable technologies that can address some of our most significant problems, and we can begin to slide down the steep and sustained learning curves that many of the technologies exhibit, we may have a realistic shot at saving ourselves! Once we hit economic breakeven with these new technologies, the economic drivers will provide a steady tailwind that will accelerate adoption of sustainable solutions on a global scale. It is imperative that we look at scaling challenges holistically, so that the law of unanticipated consequences does not once again render our gains null and void.

Conclusions

While this chapter may feel like it was a bit of a diversion, we believe that the preceding discussion directly talks to a major gap that exists in how we (at least in the US) translate deep-domain discoveries and disruptive innovations to market and to impact at scale. We believe also that unless we can drastically shorten the time and reduce the cost of this translation, many key technologies that hold the possible solution for the many challenges we are facing will not see the light of day. Nowhere is this more important than in the fields of energy and climate change. These hidden technology gems possibly hold the key to a prosperous and sustainable society, but only if we can achieve this over the next 10–20 years.

We have seen that just a few disruptive scientific advances—semiconductors, new materials, processing of materials at the nanometer scale, and digitalization have radically changed our world—starting from the 1950s, but with dramatic acceleration over the last 20 years. These few technologies have completely transformed our lives, particularly in the critical field of energy. We have seen in this chapter that these technologies may only represent the tip of the iceberg, and that there are likely thousands of other promising technologies that may be stuck in the "innovation logjam," with hundreds of more discoveries occurring every year as knowledge generation also accelerates. As we struggle to save humanity from the potentially devastating impact of climate change, we have to wonder whether there are other brand-new solutions that we cannot even imagine, and which may feel like magic today, that are waiting in the wings to be discovered and that can have impact at scale. Of equal importance is the question of whether these discoveries can be commercialized at scale in a timely manner to mitigate the worst impact and to help us adapt. This may have seemed to be impossible in the past, but *the last 20 years has taught us that new technologies can emerge from nowhere and disrupt large mature industry sectors, including, as we have seen, in the energy sector.*

A random "collision" between an entrepreneur, investor, and technology specialist is not the ideal way to commercialize high-potential deep tech. We have described a process here where the first stage of major de-risking can occur within the research center itself, one of the few places where appropriate deep-domain competencies exist. We have seen that new mechanisms may be needed to streamline how this occurs, whether in a research university or in a national lab. This includes a process of in-situ de-risking and incubation based on a holistic storyboarding process that uncovers and helps to frame the fundamental questions, the answers to which will shape the trajectory and the timeline for technology validation and scaling. Additional mechanisms for financial flexibility and sustainability for the research center are key elements for being able to move fast. If a technology is sufficiently de-risked, it becomes more accessible for further funding and translation—either by an established company or through a startup. We feel that the aforesaid commercialization efforts can dramatically impact the seemingly "impossible to solve" problems of environment, decarbonization, water, and anthropogenic climate change.

In addition to energy and decarbonization, which this book is focused on, there are a wide variety of issues that impact us that science and technology have the ability to address. These issues range from access to clean water, cooking, desalination for potable water, decarbonization and electrification of industry, redefining agriculture, recycling of everything and circular economy, societal resiliency, complex intelligent systems, and making all systems sustainable. Solving these hard problems quickly can help address some of the biggest challenges that face humanity.

References

1. Internal white paper prepared by the authors in 2015 towards formation of new structure of the 'Center for Distributed Energy' at Georgia Institute of Technology, supported by Georgia Research Alliance to drive necessary process changes (Available upon request).
2. Schaufeld, J. (2015). *Commercializing innovation, 'Turning technology breakthroughs into products'*. Apress.
3. Bayh-Dole Act. (1980). https://en.wikipedia.org/wiki/Bayh%E2%80%93Dole_Act
4. Sharma, S., & Meyer, K. (2019). *Industrializing innovation—The next revolution*. Springer.
5. AAAS (American Association for the Advancement of Sci). (2018). *Historical trends in federal R&D, budget, and policy program*. https://www.aaas.org/page/historical-trends-federal-rd
6. Brookings Institute US R&D: A troubled enterprise – Brookings Institute Report. https://www.brookings.edu/blog/the-avenue/2015/05/28/u-s-rd-a-troubled-enterprise/
7. Long, Heather. (2016). *A historic low in USA startups*. CNN Money report summarizes statistical data and primary reasons. http://money.cnn.com/2016/09/08/news/economy/us-startups-near-40-year-low/index.html
8. Price Waterhouse Coopers Corporate. (2017). *Spending hits record high but many executives have concerns*. https://www.pwc.com/us/en/press-releases/2017/corporate-rd-spending-hits-record-highs-for-the-top-1000.html. A PwC Research Report.
9. University of Pittsburgh Innovation Institute. (2018). *Bayh-Dole act at a glance*. https://www.innovation.pitt.edu/resource/bayh-dole-act-at-a-glance/

10. Midtown Tech Square. (2023). *Story of a Leading Innovation Ecosystem next to Georgia Tech*. Atlanta Business Chronicle.
11. Source IEA Information. https://www.iea.org/data-and-statistics/charts/venture-capital-investment-in-energy-start-ups-by-technology-area-for-early-stage
12. Jenson, R. A. (2011). Startup firms from research in US Universities. In *Handbook of Research on Innovation and Entrepreneurship 2011* (pp. 273–287). Library of Congress Control Number: 2010927657). Edgar Elgar Publishing, Inc.
13. Pitch Book Data – Mar 2023.
14. Paytas, J. (2011) *How many startups can a university support?* https://fourtheconomy.com/how-many-startups-can-university-research-support/ 'Fourth Economy'.
15. Lewis, H. L. (2019) Climate – Lapham's quarterly – XII, 4, American Agora Foundation

Further Reading

Brown William, H., Malveau Raphael, C., McCormick, I. I. I., Hays, W., & Mowbray, T. J. (1998). *"Anti patterns"—Refactoring projects in crisis*. Wiley.

Drucker, P. F. (1956). *Innovation and entrepreneruship*. Harper Collins Edition in 1993.

Drucker, P. F. (1998). The discipline of innovation. *Harvard Business Review, 76*(6), 149–157.

Dyer, J., Gregersen, H., & Christensen, C. M. (2011). *The Innovator's Dna—Mastering the five skills of disruptive innovators*. Harvard Business Review Press.

Lewis, H. L. (2019). Climate – Lapham's quarterly – 12, 4, American Agora Foundation

Sharma, S., & Meyer, K. (2019). *Industrializing innovation—The next revolution*. Springer Nature.

Aligning Innovation, Economics, and Decarbonization

8

Accelerating the Process of Tech to Market (T2M)

We are poised at a delicate point in time, on the cusp of a transformation in our energy infrastructure, where decisions and investments we make over the next few years will impact what kind of future world we build for ourselves and our children: a sustainable and equitable world of plenty, or a dystopian world fraught with endemic shortages and continuing geopolitical mayhem. We have seen that twentieth-century technologies and the incumbent companies and organizations that profit from them may be unable to address the core issues that are at the heart of this transformation. The last 20 years have also shown that some twenty-first-century technologies have been able to rapidly establish themselves at scale, even in the face of extreme pushback from the incumbents, and now hold the promise of transforming our lives for the better.

The last 20 years have also shown that some twenty-first-century technologies have been able to rapidly establish themselves at scale, even in the face of extreme pushback from the incumbents, and now hold the promise of transforming our lives for the better.

As we have seen, at a high level, this is predicated on an improved understanding and deeper knowledge of materials and the ability to manipulate them at the microscopic level. It is also based on converting passive mechanical assets into dynamic intelligent assets that can work together to improve value delivered. At the heart of this is the explosion of scientific knowledge that has occurred since World War II, a process that continues and is now global and unstoppable. We have explored how a few of these technologies, first semiconductors and integrated circuits, and more recently advanced batteries, are transforming everything—from our energy systems

D. Divan, S. Sharma, *ENERGY 2040*,
https://doi.org/10.1007/978-3-031-49417-8_8

to how we live, all in the short span of 20 years or so. As we have reiterated, we believe this is a result of exponential technologies that exhibit steep and sustained learning rates over decades, where performance continuously improves even as price declines. We have seen that even the few of these technologies that have made it to market have already had such a large impact—raising the question about whether what we are seeing is the tip of the iceberg, and if there are other similar exponential technologies waiting in the wings that could be equally transformative. If so, how many years will it take for these technologies to become commercially viable and to have impact?

We have also examined the ecosystem for taking scientific knowledge and converting that into products and services that provide economic value. This process has changed dramatically since the late nineteenth and early twentieth centuries, with an increasing dependence on deep-domain knowledge, scientific equipment as well as advanced modeling, simulation, and manufacturing, all used to develop and demonstrate the new functionality that underlies the scientific discovery. This knowledge has to be translated into devices or processes that can provide value and meet specific market requirements, including performance, manufacturability, cost, reliability, and life, before the translation to a commercialization scaling partner can be considered completed. We have seen that as industry, particularly in the US, has moved away from long-term R&D, a significant gap has emerged in this translation process.

The good news is that there is an abundance of new science that has the potential to holistically address key societal needs, and to realize the steep and sustained learning rates that are characteristic of desirable exponential technologies. Not every one of these will be successful, and not every one of them will scale. But we need a process where we can, at an early stage and at low cost, examine the potential viability of these technologies, addressing the key questions that can help to reduce innovation, technology, manufacturing, and market risk, allowing the creation of holistic solutions that can scale rapidly without getting mired in unanticipated consequences. As the previous chapter has shown, university research centers and national labs that do the scientific research and have the requisite deep domain knowledge can potentially undertake the tasks of de-risking, incubation, and acceleration. However, we have also seen that they frequently do not have the skills or the incentives to analyze and understand—how these new disruptive solutions can be unlocked, how they may potentially impact or disrupt an existing market, and what the technological and commercial roadmap could be for reaching scale.

As discussed in the previous chapter, at the Georgia Tech Center for Distributed Energy, we have been using Innovation-by-Design (see Chap. 7) as a process that allows the science and research to be done with rigor and independence, while still creating alignment with desired outcomes through a process we call "holistic storyboarding," where we first look at the issues very broadly, and keep asking key questions about the potential for translation and scaling very early in the problem definition process. If the outcomes from the early stage research are positioned well, the next step can involve the rapid build of a minimum viable prototype (MVP) and the validation of key questions that would allow the technology to spin-out quickly and with much lower risk. We have seen that such an approach can shrink time to market from 10–20 years to 3–4 years and can significantly reduce the amount of high-risk capital needed for commercialization.

Such an approach can also leverage resources because it is well aligned with how the government commits its funding for R&D and can ensure that outcomes are well aligned with societal objectives, as is often required for the expenditure of government funding. A complete "tech-transfer" package is needed that includes—intellectual property (patents) access, MVP with key functionality along with validation and documentation, a preliminary plan showing how the proposed solutions map over to the market and whether the technology is incremental for existing market leaders or is disruptive, and which includes results from a possible field trial/pilot. Such a "tech-transfer" package can create high value, allowing industry to decide if this is a good fit and for entrepreneurs and investors to decide on moving forward.

In any case, whether we look at the technologies that are already in play in the market or waiting in the wings in a research lab or university, it is clear that there is an abundance of new exponential technologies which have the potential to transform our lives, possibly even over the next 20 years, if done right! What are these technologies, and if successful, what impact could they have? With so much money at stake, and so many companies competing for success, how, if at all, can we make this transition relatively smooth and as painless as possible?

> *Whether we look at the technologies that are already in play in the market or waiting in the wings in a research lab or university, it is clear that there is an abundance of new exponential technologies which have the potential to transform our lives, possibly even over the next 20 years!*

Abundant Exponential Technologies

This is an exciting time as human ingenuity and innovation allow us to leverage our deeper understanding of materials and complex intelligent systems to find solutions to problems that seemed intractable even a short 20 years ago. Although advances are being made in every field, from medicine to agriculture to computing and communications, we will focus on those areas that meaningfully intersect with the world of sustainable energy in terms of generation, delivery, and its use. We are looking for technologies that have the potential for accelerated societal impact at scale, even when viewed with a holistic and sustainability lens. We need to be mindful of unintended consequences, aware that economics, policy, equity, and relief from geopolitical hegemony in raw materials should be a part of the discussion. We need a specific goal of not only protecting the environment, but also correcting the damage that we have done to it. We need a broad palette of solutions to tackle such a broad range of issues and are fortunate that we have thousands of our best minds globally that are focused on developing and deploying such solutions.

There are literally hundreds of new developments in each field, and it would be impossible to cover all of them, or to predict which ones will become successful. The goal here is to show representative ideas and the underlying thinking as to why such capability could be valuable in our journey, and to show that we can get to alignment between the economics and sustainability—a major element of which, for energy, is decarbonization and circular economy. As we discuss some of these new ideas (in no specific order), we will also discuss some of the opportunities and challenges that should be considered.

Photovoltaics

Clearly the biggest ongoing disruption is occurring around PV solar energy (wind, as we have discussed, may be reaching limits to how long it can continue along the steep learning curve, although it should be noted that it is already well below the cost of traditional fossil-based generation and can keep gaining market share). There are many new technologies being developed for the next generation of solar cells, including cells based on perovskites as well as organic solar cells. Perovskites are thin film devices built with layers of materials that can be deposited on an inexpensive substrate using printing, coating, or vacuum deposition techniques. By converting a larger portion of the incoming solar energy, they can also achieve higher efficiencies than silicon, with ~25% efficiency numbers already reported. Tandem cells with perovskite on silicon have also been built with efficiency >33%, with even higher levels possible. Even lower cost may be achieved with organic solar cells, where abundant non-toxic organic materials can even be sprayed on to a substrate such as paper or glass to realize ultra-low cost.

Over the last ten years, organic cell efficiencies have increased from ~3% to over 17%, with continued increases anticipated. Additional developments such as building-integrated PV and "solar paint" point to ongoing efforts to continue to reduce cost of solar energy available. Over the next 20 years and more, it seems highly plausible that photovoltaics will continue to progress along a steep learning curve, potentially bringing the cost down to a point where solar energy becomes abundant and inexpensive. An aspirational levelized cost of energy (LCOE) goal of <$10/MWh for utility scale plants and a vision of solar energy as abundant and ubiquitous definitely seems achievable over the next 20, maybe even 10 years.

Even as the potential of abundant, almost free, solar energy dazzles our imagination, we also need to understand the challenges that lie ahead. The manufacturability of high-efficiency large-area thin film cells and ensuring that they can last for the target 20+ year life, presents major challenges (whether they need to last for 20 years with steep learning curves is a different discussion altogether). With an exploding market comes the question of a circular economy, something that can only come from policy and mandates. The research community, systematically assisted through grants from DOE and other institutions in the US and even more aggressive support in Europe, China, and other nations, is continuously moving the ball forward on all these issues. Because the PV disruption, propelled by fast moving technologies, is now already underway, there is less resistance from manufacturers who are now actively looking for new technologies to make sure they can maintain or get into a leadership position in a proven and fast-growing market.

One major issue that we have discussed earlier is that much of the manufacturing of solar cells and panels has moved to China, with few US manufacturers remaining. This results in geopolitical vulnerabilities of a different kind than for oil, but also creates situations that need to be addressed. The good news is that as the cost of PV cells continues to decrease, the cost to transport the panels and to assemble them at site will become a larger part of the overall cost and will make local manufacturing more favorable. Also, because this technology is moving fast, one can always build PV panels using different or slightly older technologies (with say a 1% efficiency penalty), and still come out ahead because PV is so much cheaper than other alternatives. *This dramatically reduces the geopolitical vulnerability, because unlike for oil, we always have other viable options*—which should encourage offshore solutions to remain competitive and to reduce geopolitical vulnerabilities. We have already seen what happens when we sacrifice national capabilities and security for a few pennies more in profits! We don't want to go there again.

Energy Storage

We have discussed the importance of energy storage for both mobile and stationary applications. The availability of cost-effective advanced Lithium-Ion batteries started the disruption in transportation that we are living through. Early adoption was driven by battery chemistries and formats, such as the 18650 cell that was developed for the electronics industry but was then used for EVs—a non-optimal step but one that allowed rapid scaling. As the EV space has become more established and more competitive, we are seeing a natural migration to better battery chemistries and form factors.

Existing Li-Ion chemistries include NMC (lithium, nickel, manganese, and cobalt-oxide cathode) and LFP (lithium iron phosphate cathode) with a graphite anode, which have energy densities of around 250 Wh/kg and a life of 1500–3000 charge/discharge cycles. The use of silicon anodes can increase storage capacity by 20–40%, while the advent of solid-state batteries could allow faster charge/discharge and reduce fire risk by allowing batteries to operate at higher temperatures. New battery chemistries such as Lithium-sulfur (Li-S) and Li-air offer even higher energy densities, with a theoretical density for Li-air of 13,000 Wh/kg, 50× of what we can achieve today with Li-Ion. Finally, different batteries, such as sodium-ion batteries work with abundant materials and are now being offered commercially, while gold nanowire batteries hold the promise of over 200,000 charge/discharge cycles—virtually infinite life! Again, as in the case of PV solar cells, there are substantial markets with strong growth that are being served by batteries today at price and performance points that are already competitive in many high-volume use cases, with further broadening of the market as prices continue to decline. This demand is fueling further research to address key gaps in performance, price, and safety, with a systematic progression and a sustained and steep learning rate which ensures that advanced batteries will be able to meet the needs of the EV industry for decades to come. Battery demand for EVs alone is projected to increase to 1200 GWh by 2040 with battery prices of $76/kWh and lower.

As we have seen before, batteries for 2–8 h of grid storage appear to represent a much smaller market than EVs at this time. Automotive batteries are available at low-cost and are being used for grid storage, with good return for several use cases [1]. It is interesting to reiterate that 125 million EVs in the US by 2035 would represent 9000 GWh of energy storage (versus 600 GWh projected globally as the demand for grid storage) and 25,000 GW of inverter capacity (versus 1000 GW of US generation capacity today), that would frequently connect to the grid. Being able to realize value from the EVs using vehicle to grid (V2G), vehicle to home (V2H), and other applications could provide great value in terms of grid support, emergency backup power, and community resiliency. The need for high levels of energy storage is spurring new innovation, including companies such as Form Energy with iron-air batteries that are promising long duration energy storage at ~$20/kWh, as well as gravity batteries that move massive amounts of material to convert potential energy to electricity.

The additional value that comes from adding storage to DERs is being recognized by regulators, and an increasing number of states in the US are approving hybrid PV plus storage projects, with 65 GW/260 GWh expected over the next few years, representing >80% of new solar installations. These projects will also be riding the twin learning rates of PV and batteries (not to forget, power converters needed for both), with decreasing prices and longer storage times for the foreseeable future. Such hybrid plants may hold the key to meeting rapidly changing demand for high-peak loads such as EV charging, particularly in urban centers, as well as for providing a resilient power system for vulnerable communities.

There is continuing interest in the idea of second-life batteries, where EV batteries that are at "end of life"—typically 90% of the range at the start—can be used for grid storage. While this seems like an interesting idea, the long-term numbers seem challenging. EV battery life is being projected at 100,000 to 500,000 miles, or 10–20+ years, and improving. The residual cost of the battery plus the cost of testing, shipping, and repackaging the battery, and then mixing it with other car batteries, possibly from different manufacturers and of varied ages, seem to be expensive and difficult-to-scale propositions. In the meantime, the cost of a battery pack with the same original rating could have decreased by 50–80% due to the learning rate, improving returns for the specific targeted grid-storage use-case. During the same period, battery recycling would have progressed to the point where recyclers may pay for the raw materials in the battery, creating a simpler supply chain, and taking care of the end-of-life issues for batteries. Lead acid batteries today recycle 98% of the very toxic lead that is in the battery—there is no reason why Lithium cannot get to the same level of recycling efficacy over the next 10–20 years, and why they should not be required to do so under new regulatory fiat.

Zero Emission Dispatchable Generation

As we look beyond short-term storage, there is need for mid-term (1 day to 1 month) and long-term (many months) storage as well. These are critical for grid operations and may also be important for high energy-need applications such as heavy trucks, ships, locomotives, and aircraft. For the grid, the need for long-term dispatchable generation with zero greenhouse gas (GHG) emissions can be filled in many ways, including with hydropower, geothermal power, nuclear power through the deployment of small modular reactors and micro reactors, as well as the use of green hydrogen, or blue hydrogen from natural gas, but with permanent carbon sequestration, all of which have been discussed in a previous chapter. Geothermal energy is limited by location but may be an attractive option where available. For fission-based nuclear energy, spent fuel management, proliferation risk, danger of accidents and community pushback continue to deter wide deployment. Nuclear fusion, on the other hand, seems attractive (particularly with the recent announcement of the first energy breakeven reaction) but remains a longer-term goal. While much progress has been made, the first viable commercial scale fusion plant is not feasible before 2040, the period under consideration here.

We have already seen that hydropower provides a very good balancing resource to offset the variability of DERs, including the ability to continuously operate for weeks and months when required. Converting existing hydro plants to pumped hydro offers another opportunity to provide the flexibility that grid operators need to balance the system. Wherever possible, existing hydro plants could be converted to hybrid plants, where they are also coupled with collocated PV solar or wind resources. Such plants can provide the needed energy >90% of the time from the DERs using hydropower primarily as a dynamic balancing resource, preserving the water for human consumption and for environmental purposes, while still retaining the ability to use it for continuous dispatchable power generation in an emergency. The grid-integration costs are also low because the substations and transmission needed to connect the PV/wind to the grid already exist. Similarly, collocating PV with an existing depreciated gas generation plant can eliminate the need for new transmission, cut emissions by 90%, and still realize the reliability and long-term backup the grid needs. Such a hybrid approach can provide a good transition strategy.

However, by no analysis does hydropower exhibit the characteristics of a steep learning rate technology. Its appeal is primarily in the case of an existing depreciated asset which can be deployed at minimal additional cost to provide new value to help balance the grid. A similar possibility exists where two existing reservoirs can be interconnected with pipes to realize pumped hydro functionality. Interesting solutions where floating PV is deployed in conjunction with hydropower shows dramatic benefits, where one can ride the steep learning curve of PV for the bulk of the energy delivered, but also retain the balancing and long-term backup property of the hydro asset. This allows an integration of the declining price of PV solar and the already low cost of an already depreciated storage asset. It is, however, geographically very limited. Interesting new solutions, such as creating a "lake in the ocean" show novel approaches are possible to expand pumped-hydro benefits beyond traditional hydropower locations but have not been proven so far.

Of all these solutions, perhaps the one that has the potential for a steep learning rate and shows great promise is green hydrogen. Major projects underway for tying PV solar generation with hydrogen electrolysis show a pathway for achieving a cost of less than \$1/kg for green hydrogen as needed to unlock what could be a \$100B/year market by 2030. Key innovations for reducing capital cost of the electrolyzer from \$700/kW today to under \$200/kW in a few years include replacing noble metal catalysts with cheaper, more readily available minerals, photo electrolysis of water, hydrogen generation directly from seawater, solid oxide electrolysis cells (SOEC), and reversible fuel cells that can convert electricity to hydrogen and vice versa in the same device, thus greatly simplifying the overall system. Taken in conjunction with declining PV solar energy prices, the path to \$1/kg for green hydrogen seems clear, as attested to by over 250 GW of global green hydrogen projects already in the pipeline. Green hydrogen can not only serve as a generation resource for electrical power but can also substitute fossil fuel-based heating for industrial processes, commercial buildings, and heavy transportation.

The other approach using blue hydrogen with permanent carbon sequestration is to take natural gas (which is plentiful) and to use microwave plasma pyrolysis or other similar technique to generate hydrogen and solid carbon in the form of graphene, or another organic feedstock. The carbon-chain products permanently sequester the carbon and enable use of hydrogen as a fuel with no carbon emissions (fuels produced with this technique still result in carbon emissions). As discussed earlier, because energy is needed to break the hydrogen atoms free, the net energy realized from this multistep process is not so favorable. However, the value of the carbon products, such as graphene, can allow a much lower cost point for the carbon-emission free hydrogen. It should also be noted that every ton of hydrogen generates 17–20 tons of carbon-based products, which may provide high economic value initially, but can become a challenge once the market for the graphene or carbon products becomes saturated. On the other hand, using the carbon in a carbon capture and utilization (CCU) process to generate much needed chemicals and raw materials can provide longer-term sustainability. Additional questions related to hydrogen leakage into the atmosphere, including GHG impact and danger of potential explosions, also need to be addressed.

Another new technology, Direct Air Capture, is being used to capture the CO_2 from smokestacks of conventional generation plants, dissolving the CO_2 in a solvent, and subsequently releasing it through the application of heat or precipitating it out of the solution. A second related approach involves the extraction of CO_2 directly from air, using a similar process. This CO_2 can then be permanently sequestered in underground caverns by mineralizing it with basalt rock over a short period of a few years. Direct Air Capture is generating a lot of interest and billions of dollars in funding from government and investors with a target of getting to ~1 G-ton of carbon removed by 2050. The process needs a lot of energy to operate—making low-cost resources such as geothermal and low-cost PV critically important for achieving the goals. DAC offers a remediation process for actually reducing the concentration of CO_2 in the atmosphere, and as such is an important tool in our toolkit. However, it offers little commercial value other than the capture and sequestration of CO_2 and will possibly require a carbon tax, carbon trading, or mandate governmental funding for successful rollout and scaling of this technology. Philanthropic funding, critical in the early years, may not be sufficient to achieve commercial viability and scale.

Electrification of Everything

We have discussed at great length the electrification of on-road vehicles, including scooters, tuk-tuks, automobiles, trucks, buses, and electric semis—and the opportunities and challenges that this scaling presents. In the US, transportation today is Internal Combustion Engine (ICE) based and represents 6% of the energy used, in this case delivered to the wheels, but also accounts for 33% of the total wasted energy and 27% of total US CO_2 emissions. We have also seen that EVs are on a trajectory for continued performance improvements and price declines, enabling a transformation to an electrified fleet. Similarly, for heavy vehicles, off-road vehicles, ships and aircraft, there is substantial global effort to reduce the carbon footprint through electrification, possibly using a mix of batteries and hydrogen. EVs are following a steep learning curve and will continue to increase market penetration over the next 20 years. The challenge will possibly come from the need for a ubiquitous charging infrastructure, very likely including a large number of fast-charging access points. We have seen that the issues of peak demand and the need to build new generation at the grid edge as well as new T&D infrastructure may prove very challenging in the time that is available. A more distributed grid-edge strategy could address some of these issues but faces regulatory challenges for deployment across all states in the US.

While we have focused attention on transportation, the industrial and commercial sectors also consume a significant amount of energy and are responsible for 31% of GHG emissions. Cement, iron and steel, and chemicals are the largest GHG emitters globally, accounting for almost 6 Gt/year in CO_2 emissions. In the US, out of a total energy use of ~100 quads in 2019, the industrial sector used 22 quads of energy from petroleum, natural gas, and biomass, mainly for process heat, and only used 3.25 quads in electrical energy. From the 2019 data on US energy consumption, we can see that a straightforward conversion of the heat needed for industrial processes to electricity will require around 10 quads of new electrical energy (given 45% efficiency), representing a 40% increase in the electricity generated today. In some cases, this could also be met with green hydrogen, or through blue hydrogen with permanent sequestration of the CO_2 generated. Such a major shift in energy resources, if driven by a need to reduce CO_2 emissions and done indiscriminately, could be very expensive and difficult to justify. On the other hand, if the justification is economic in nature, driven by the energy savings that the manufacturing plant would realize, competitive forces would drive the conversion. The prospect of low-cost dispatchable solar and/or green hydrogen at <$10/MWh is very attractive and could drive such a transformation. Incentives may help to accelerate the process.

On the other hand, it is not clear that using the processes that were developed, in some cases hundreds of years ago, provides the best approach today for the manufacturing of everything from cement and fertilizer to metals and feedstock chemicals. For instance, the Bessemer process for making steel out of iron ore dates back to 1000 BC, and is still used today, accounting for 2 Gt/year of CO_2 emissions. Newer electrical processes, such as those based on molten oxide electrolysis, can directly convert the oxide to metal in one step, with no carbon involved and with the ability to operate from renewable resources with their steep learning rates [2]. A high level of modularity also allows scaling of a plant as needed. Large scale users of steel are already responding, with companies such as Volvo committing to build cars from zero carbon steel.

Similarly, ethylene exemplifies how chemicals also provide emerging opportunities for electrification and decarbonization, although not without challenges. In the case of chemicals, ethylene is a key raw material that has been called "the World's Most Important Chemical" and is produced at petrochemical refineries using the cracking process. It can subsequently be converted into a vast number of products, ranging from plastics, PVC, Styrofoam, and ethylene glycol, and accounts for 10% of the CO_2 emissions by the entire chemicals industry. McKinsey has shown that electrification would be viable in greenfield plant electrification if renewable energy was available at $25/MWh. For electrification of brownfield plants, the renewable energy rate would need to be at $15/MWh. As we have seen, these numbers are viable today, and will be decreasing even more over the next few years, allowing electrification of increasing parts of the chemical industry. Once again, we see that economics and decarbonization can go hand-in-hand.

Looking at the factory of the future, we also see opportunities for modularization, in-situ manufacturing, distributed manufacturing enabled by 3D printing, and other advances in additive manufacturing techniques that will be transformative. Such advances can mass produce products in a region (as opposed to centrally), thus significantly reducing the burden on supply chains (logistics and transportation). This can also promote regional economic growth and promote designs that are more compatible with local resources. Potentially, every community (even every home) could be a node for a next generation clean factory, moving us from a polluting centralized model to a clean distributed model while achieving the same outcomes or better.

The science of manufacturing is providing new options to flexibly manufacture everything from food, medicines, textiles, concrete, car parts to new types of materials by using additive manufacturing principles, essentially building these materials and structures layer by layer. By using locally available raw materials, we will also reduce geopolitical tensions, and reduce the energy and monetary cost of transportation. In the case of food, we are already seeing a reinvention of how familiar food products can be made, while reducing the GHG impact and resource use. All these innovative new technologies sound like science fiction, but are reality today, and have a good chance of going mainstream by 2040.

All these innovative new energy technologies sound like science fiction, but are reality today, and have a good chance of going mainstream by 2040.

Commercial and Residential Applications

Finally, we should not neglect the need for novel solutions in the commercial and residential space—particularly for solutions that mitigate or adapt to climate change and global warming. Solutions needed include personal transportation (including personal mobility and mass transportation), cooking, water, cooling/heating of buildings and resiliency that allows us to continue to live our lives in the face of HILF events. This is also very challenging because the life of buildings tends to be 30–60 years, which makes it challenging to upgrade the infrastructure. There are a wide range of new technologies aimed at buildings including building-integrated PV, thermal energy storage using structural elements, and lighting and cooling that is based on occupancy, and 3-D printing methods to reduce build cost and time for new buildings. Cooking lends itself to electrification and could be addressed by the availability of low-cost electricity, as well as by city regulations that are increasingly forbidding gas-based cooking for new houses.

The biggest challenge is possibly air conditioning and space cooling. Along with low-cost energy, this also requires low ozone depleting potential (ODP) refrigerants. New technologies such as metal organic framework (MOF) use interstitial spaces to manage humidity, reducing the energy need for a given level of cooling by 75%. Similarly, another new approach, also using novel materials, applies a hydrophobic (water repellant) ceramic to a highly water-absorbent ceramic to create a very efficient heat exchanger. Even more exotic cooling systems that work on thermoacoustic and thermoelectric principles are available, already showing up in the form of inexpensive beer coolers! Even though these technologies are somewhat less efficient than conventional air conditioners and coolers, they use no refrigerants and may have a very attractive future in a world where PV solar energy can be almost free when there is abundance. With heating ventilation and cooling (HVAC) units set to increase from 1.6 billion today to over 5 billion by 2030, this is a very serious problem that needs "out-of-the-box" solutions.

As these new solutions, that ride on low-cost PV solar energy, become available, emerging markets stand to gain the most as lower cost energy and new "smart" functionality will allow them to leapfrog the developed nations and achieve a high standard of living without the high level of energy consumption that is needed today to sustain it.

Autonomous Intelligent Energy Devices

While we have talked about technologies that perform basic energy-related functions, we should also discuss the role of decentralized computing, intelligent devices at the edge that can sense and effect change autonomously without low-latency central coordination, as well as integration of artificial-intelligence and machine-learning to improve device and system level performance, and the impact that such devices can have on the energy system. Intelligence and autonomous dynamic control capability are at the heart of everything from autonomous vehicles to inverter based resources (IBR)-rich grids and microgrids.

Autonomous vehicles have grabbed the lion's share of attention in the media, seducing us with a Jetsons-like image of relaxing or sleeping in a car while it safely gets us to where we are going. Truly autonomous driving in a variety of end-use conditions, ranging from a crowded bazar in India, to situations where the car has to decide in real-time between two very bad choices, continue to be problematic in terms of sensing, decision-making, action, safety, and then liability. In open unrestricted systems, these actions have to be taken without relying on an ability to communicate with adjacent devices (vehicles), and to be able to operate with latencies or cybersecurity issues with communications to the cloud. Very smart people and well-funded organizations are tackling this problem, and maybe they will succeed, or more likely will introduce solutions that work in specific restricted conditions or environments, or by acting as more of an advanced driver-assist system. The challenge for autonomous driving of course is in the case where the conditions change, and the car is not aware of it or cannot sense it—and the driver is resting! Who is liable in such a circumstance? In any case, while autonomous vehicles could have some impact on the energy system, it will likely be minimal, especially if the cost of energy is as low as we think it could be.

From an energy system perspective, what is perhaps more important for autonomous devices is the ability to operate and control millions of grid-edge devices that will be connected to and will constitute the future grid ecosystem. We have discussed at great length the challenges of managing a future grid using today's centralized grid paradigm, where it is assumed that every element on the system is known and controlled on the basis of off-line state estimation and optimization, followed by communication of dispatch signals to the generators. Unlike for the internet, which only manages data flows and does not crash when there is latency or delay, we have shown how the real-time physical interactions between the various grid components makes centralized control of millions of autonomous inverter devices very challenging. As we have discussed earlier, the bulk power system is seeing increasing levels of IBR penetration, with challenges in terms of grid stability and controllability. Every inverter today is controlled by proprietary software which is custom-designed and generally complies with standards and grid codes that were in place when the device was designed. As a result, these standards are generally reactive in nature, lagging by 6–10 years and typically address issues that were observed and considered important in the past.

What is needed is a forward-looking strategy that anticipates the issues that will become a challenge as the penetration of IBRs on the system increases. This should then be used to formulate rules that will allow these inverters to "collaborate" with each other in real-time using decentralized control techniques that are not impacted by latencies or failure in communications, or by cyberattacks. Such devices should continue to operate in the face of normal, abnormal, and "fault" conditions, without tripping and bringing the entire system down. These inverters will need to be intelligent and able to operate autonomously, under the command of the grid operator when needed, but should also be able to autonomously operate in islanded or microgrid modes [3]. This is a very challenging problem, solving which can have a big impact on how the grid of the future will operate.

Researchers and companies are working on developing the technology and products that can realize such a flexible power-electronics-based ad hoc and fractal grid that can be built from the top-down or bottom-up, using highly interoperable modular plug-and-play devices, which follow "collaboration" rules, can be scaled as needed, can work with today's grid, or can be a building block for the future grid [4]. Locating such devices within the grid and at the grid edge, and integrating them with energy generation and storage resources can create a future grid that also has all the features and functionality we need today, and can be an integral part of today's grid operations. These devices can address issues of peak demand, load balancing, "Duck-Curve," frequency regulation, inertial support, and reactive power injection to meet current grid connection needs. On the other hand, they can also operate as an ad hoc decentralized microgrid upon receiving a command from the grid operator, or autonomously if the main grid is down. No such autonomous devices are commercially available today but are now starting to become available in labs such as the Georgia Tech Center for Distributed Energy (CDE) [4].

Stand-alone applications that do not need to work in an interconnected manner (e.g., cars), can scale rapidly because they operate as individual devices and do not need to autonomously operate together in a physically linked and tightly coordinated manner. Conversely, grid-connected power-converter devices need to be designed and certified to operate together to sustain the grid, and to do so without interfering with each other or with the system. Today, this requires a long and convoluted process including a grid interconnection study, and the myriad and fragmented regulations that they need to comply with—most of which do not fully address the future emergent requirements of an IBR-rich grid anyway. There are several companies that are offering solutions that address some of these requirements, but an overall scalable solution is lacking, as is a forward-leaning policy that incentivizes the creation and deployment of such flexible solutions. We are also lacking a clear and well-articulated vision of what we want the future system to be capable of.

What we have discussed above are examples of scientific initiatives that are underway globally, many of which have the potential to completely change the way we live our daily lives and could redefine our relationship with energy. We have also highlighted some of the gaps that will increase confusion and might cause us to move toward alternate scenarios that are suboptimal. We do see that there is a lot of technology that is being developed, technology which has the potential to impact what our energy future could look like. But can these technologies get to scale in a short time span of 20 years? And if so, who is doing it?

Who Is Taking These Technologies to Market?

The rapid disruptions underway have triggered a tsunami of activity at global scale—in terms of response by industry, as well as by investors, VCs, entrepreneurs, researchers, and scientists, who are all looking to create and ride the next wave of disruption. The automotive sector has clearly embraced and is now driving the change in their area. Solar, batteries, EVs, and power electronics are also riding steep learning curves and seeing tremendous market growth. Supply chains for these emerging industries, ranging from Lithium and copper to semiconductors, are ramping up to meet incipient demand. Established incumbent industries are positioning tired old products from yesterday as green-washed new products for tomorrow or are pushing toward a future that incrementally builds on investments they have already made over the last many decades. Oil companies and electric utilities are making high-level pronouncements about how they are also ready for a future where their existing core competencies and business models may be inadequate to meet the new capabilities needed.

Investors are seeing the opportunity provided by the intersection of fast-moving technologies and the disruption that it will create and are investing very heavily in new companies, with a record $900B committed just over the last three years. Billionaires of the world under Bill Gates's leadership have come together to systematically transform industries that are today's major CO_2 emitters. One example of a major initiative is Breakthrough Energy (BE), which is the umbrella organization focused on reducing global GHG emissions from 51 Gt/year to zero! Breakthrough Energy Ventures (BEV), which is the Venture Capital (VC) arm of BE, is focused on funding those companies that show a path to reducing >1 Gt/year in CO_2 emissions, all without compromising profitability.

They are targeting cement (TerraCO_2), metals (Boston Metal), hydrogen from water (H_2 Pro), Waste-to-Energy (Sierra Energy), Lithium from brine (Lilac), carbon mineralization (Heirloom, Carbon Cure), water purification (Source), recycling (Redwood Materials), fertilizers, and a host of other exciting solutions that can help us get to a sustainable and economically viable future built around net zero carbon emissions [5]. BEV is only one of thousands of VC firms that together have committed capital of >$500B and are aggressively investing to get us to a future where the forces of economics and decarbonization are aligned. These new disruptive technologies will not be introduced into the market by incumbent market leaders, but instead needs entrepreneurs and investors for success.

The current Biden administration is also focused on accelerating the transition through several major bi-partisan legislative initiatives that have injected $555 billion into building a new and green infrastructure, encouraging US manufacturing, and beginning to combat issues related to climate change. The market has giddily responded to this flurry of activity, stoking frothy evaluations, encouraging SPACs (Special Purpose Acquisition Company) to accelerate access to public funds for smaller pre-revenue companies, and anointing hundreds of "unicorn" startups with more than $1B valuations.

But at the end of the day, most of these initiatives and companies will have to be financially successful unless they are philanthropic in nature, and/or are inherently not scalable. To ensure success along a tight timeline and with limited resources, most companies will take a focused approach to solving a narrow problem within their own silo and will minimize their reliance on new risky innovations. Investors encourage such focus because they do not want to "boil the ocean," so to speak. The presumption often is that adjacent areas that can impact their success will not change by much or will be able to meet their requirements—an assumption that may be true when the pace of change is slow but is a challenge for fast growth and scaling, especially when many adjacent areas in the system are simultaneously being transformed at high speed.

What Happens When an Irresistible Force Meets an Immovable Object?

We have discussed the possible limitation that the grid-interconnection-point poses for a host of new technologies, including PV, wind, storage, and EVs—all of which will be moving very fast, at least over the next two decades. *It is the intersection of the free-wheeling and fast-growing private sector and the regulated risk-averse deliberate-pace of the electric utility sector that appears most problematic.*

> *It is the intersection of the free-wheeling and fast-growing private sector and the regulated risk-averse deliberate pace of the electric utility sector that appears most problematic.*

For instance, we have seen that large PV and wind farms need AC or HVDC transmission, which is slow or impossible to build at the level that is needed. Further, grid integration of high levels of PV and wind, and the control of high IBR penetration grids is proving challenging, leading to pushback and slowdown of new installations being approved. EV and e-semi fast-charging will pose very high peak demand on the grid, in total potentially exceeding current generation capacity in the US. Yet, the majority of the EV fast-charging equipment manufacturers are completely silent on grid issues, assuming that grid capacity will magically become available when and where it is needed. Alternatively, charging station developers are planning to put in redundant generation and energy storage at EV charging stations to reduce peak demand and to be independent of the grid if needed, which could in turn dramatically increase the cost of EV charging under current business models.

Similarly, while PV developers are busy deploying gigawatts of new farms, they are not concerned about the difficulty of integrating their energy output with the grid. In fact, their success depends on the grid accepting and paying for all the energy they can generate. They do not want any curtailment of their generation because their business model is based on maximizing generation and revenues over the 20-year project (and we know that with price declines for PV, these plants will be challenged to remain competitive over such an extended time span). Net metering for PV solar generation by residential and commercial customers, especially when done in an uncoordinated manner, can stress the distribution grid. Yet, many states require that the utilities accept the generated power and pay the PV system owners at a very favorable rate, while other states do not allow any compensation whatsoever for energy returned to the grid.

Similarly, for energy storage, it is difficult for "behind the meter" energy storage to participate in or be compensated for grid support, requiring the building of more storage resources by the utility to dynamically balance the grid. EVs store enough energy to run a home for many days (vehicle-to-home or V2H) and can in principle also support the grid (vehicle-to-grid or V2G), as has been shown by several EV manufacturers. Yet the cost of implementing V2H for resiliency and as backup generation remains extremely high, and V2G remains a dream with no widely accepted mechanisms in place to compensate EV owners. Finally, even if we had the will, the policies, and the economic models in place, we are still missing key pieces of the technology that can make such a decentralized and distributed system work seamlessly, reliably, and cost effectively.

The intersection of these two sectors has been problematic for decades. It was during the period from 2000–2010 that Sand Hill Road VCs first became enamored with energy. They could see the impact that digitalization could have on the grid and were buoyed by the Smart Grid discussions and eventually over $8B in incentives coming from the Obama administration. The entrepreneurs and VCs felt that they could follow the traditional startup development strategy—problem definition, MVP build, develop business case, identify early adopters, do a pilot, validate technology and business model, and then scale the product across multiple utilities. The need was great, and the potential payback was fast, suggesting a quick transition through the "valley of death"—the point where most startups stall out—with a vast unserved market to take over! *That proved not to be the case.*

The process for adopting new and potentially disruptive technology at scale in a utility was very challenging and followed a complex and often opaque process that took many years. Electric Power Research Institute (EPRI), as the technology focal point for utilities and the organization that had the deepest and best visibility across the entire sector, was the only organization that could have been the lead and guide on what needed to be done to manage an increasingly uncertain future where existing systems were being disrupted by fast-moving technologies. However, as desired by their utility funders, EPRI was essentially structured and functioned as a think tank, consultant, and testing organization for the utilities.

As a result, they did not have the DNA and real-world experience to actually define, develop, and build such disruptive products and solutions, to demonstrate and validate these with real-world pilots, to transfer these new technologies to manufacturing at scale, and to then guide their utility members in rolling out such new technologies (and neither did the utilities have such competency)—especially given the breathtaking pace at which this was needed. As an organization funded and controlled by the utilities, EPRI was more focused on doing what the utilities wanted—which was to guide them in the adoption of the incremental technologies and solutions that the utilities wanted, and certainly not to promote a disruption of their existing systems.

Driven by their industry members finally recognizing the need for action, EPRI is now leading the charge on an action plan for decarbonization. However, the disruptive ideas still had to come from startups or a company that wanted to disrupt the status quo. A startup that identified an opportunity to improve or impact the grid would work internally on getting to a point where they had the MVP and the business impact fully understood and would be pursuing a pilot to demonstrate the technical and financial model. Companies that managed to navigate their way to a pilot project for their new technology through discussions with the utility R&D groups and/or EPRI (necessary for new technology introduction), quickly found that it did not lead anywhere. Other utilities who expressed interest wanted their own pilots and claimed they could not learn from others.

What would have been a 3–6-month validation exercise in the commercial world was now taking five, even ten years. Companies that had to ramp up their technical, manufacturing, and sales processes and teams to get to pilot stage could not count on revenues any time soon but were burning cash as they tried to keep their teams together. Even as the pilot was successfully completed, it just meant that the next successive hurdle was then presented to them. Finally, once the approvals were all in, the actual purchase order then went into the next budget cycle. If you were really successful and were to see broad deployment, this would require your solution to be baked into the rate-base, allowing deployment at scale to finally take place, 10 years later—by which time the original solution was probably obsolete! *For companies and investors bringing new technology to the utility market, "death by pilot" became the unfortunate rule—the "valley of death" was not a valley but a deep uncharted ocean!*

For companies and investors bringing new technology to the utility market, "death by pilot" became the unfortunate rule—the 'valley of death was not a valley but a deep uncharted ocean!

The net result of this challenging intersection was that many of the new startup companies in the grid space did not succeed, and if they did, saw very slow ramp-up and as a result suffered dilution for the founders and early investors, and did not achieve their full potential. This included companies in distribution automation, smart grid, dynamic grid control, analytics, grid-edge visibility and control, and power quality. VC investment in companies looking at entering the regulated energy space, at one time the only mechanism of support for disruptive entrepreneurs, dropped from $25 billion in 2006 to $0.5 billion by 2012, despite the emergence of several strategic VCs who were aligned with major utilities and companies.

The only companies that could survive long enough and still succeed were the larger corporations and/or those fully aligned with the incremental strategies being pursued by the utilities, companies such as GE, ABB, Siemens, Mitsubishi, Itron, SEL, and S&C Electric—all long-time utility vendors, and all with significant balance sheets which the utilities wanted (a requirement that automatically excluded all startups). The highest chance of success and deployment at scale was for mature companies that were developing incremental solutions that the utility planners had specified in their long-term plans, and which had been approved by the public utility commissions—which, by definition, could not include disruptive ideas.

The coming of independent power producers (IPPs) and partial deregulation was the first major disturbance that the utility industry had faced in decades. When they saw the rise of PV solar, wind, and storage technologies, they successfully pushed back and slowed it down for more than a decade, until it was no longer possible. For the first time now, the momentum of PV, wind, storage, and EV deployment cannot be stopped. We have also seen that the electricity industry does not have all the tools it needs to manage this transition, nor can they move at the speed that is necessary. As we explained in Chap. 5, this situation leads us to an inevitable question: What happens when an "irresistible force" meets an "immovable object"? At the 2019 Energy Thought Leaders Symposium organized by our Center for Distributed Energy at Georgia Tech, a major EV manufacturer representative had a short answer to that question—"We go around it!"

The concern is that we have rapid deployment of new technologies fueled by billions of dollars that is occurring on the grid, particularly on the grid-edge, but that this pace of change is not matched by changes needed on the grid.

There is every chance that in the near term, given the highly fragmented and uncoordinated approach being taken, and the uncertainty in terms of what needs to be done, we will see events where the grid is unable to cope with abnormal "corner-case" operating situations—those with a low probability but high impact, such as extreme heat or cold, fires, and storms—all seen with increasing frequency in recent years in many parts of the US.

Regardless of whose fault it is, every hiccup along the way can delay, even temporarily derail the transition, and make it much more expensive. We think it is almost certain that the future grid transformation will occur, but it could take time. In the long run, the economics and functionality cannot be denied. On the other hand, there is a distinct possibility that the process, in the absence of a rapid and coordinated response, could turn out to be chaotic, as it has often been in the past transformations. If that happens, it could once again take the 50+ years that similar transformations have always taken in the past. By then we will have lost both battles: economics and climate change. We will also have made social outcomes much worse by increasing the gap between the haves and the have-nots. We need another path forward!

We think it is almost certain that the future grid transformation will occur, because in the long term, the economics and functionality cannot be denied.

Aligning the Forces of Change

As we have seen, this is a complex issue. We can all agree that change is coming at us like a freight train, driven by fast-moving technologies with exponential growth and declining prices. Because these technologies connect to the grid, we must also agree that something needs to be done at the grid level if potential chaos is to be avoided. But because there are so many groups that would like to maintain the status quo, and because they could lose billions of dollars in the process, the transition will be challenging.

It is difficult to handle the rapid pace of change given the limited scope and range of federal regulations, the widely varying regional regulations, and the challenge, in terms of both politics and timeline, to implement any changes needed. Overlaid on this are issues of global politics, geopolitics of raw materials and supply chains, equity, consumer preferences, regional economic and workforce development, and of course climate change and decarbonization. All these forces, simultaneously jousting for resources and acceptance, create a high level of uncertainty about how the system will evolve, over how much time, and what new solutions will be needed. How do we navigate our way through this complex maze?

We feel the best way to cut past the clutter is to elevate the discussion to a point where we can all agree on an aspirational vision of what an ideal future energy system, for us and for our children, should look like. In brief: we want an energy ecosystem that is fueled by abundant clean energy with no emissions, which is also sustainable, economical, flexible, reliable, resilient, and equitable, which meets the needs of the fast-evolving sources and loads that it serves, and which does not trample on the rights of future generations (remember, this is aspirational!).

We want an energy ecosystem that is fueled by abundant clean energy with no emissions, which is also sustainable, economical, flexible, reliable, resilient, and equitable, which meets the needs of the fast-evolving sources and loads that it serves, and which does not trample on the rights of future generations.

Such a vision can serve as our "North Star," a guide to shape our decisions and actions and help clear confusion and ambiguities. Equipped with this guide, we can derive what solutions need to be developed, solutions that are green and flexible, and can be used along a multiplicity of potential pathways, including any regional differences that may exist. We will look at the various players in the electricity sector to see what, if anything, can be done to bring about the desired alignment.

Augmenting the Existing Bulk Grid

We have seen that as DER penetration increases, as more EVs are deployed, and as electrification and hydrogen look increasingly viable, the existing grid will be increasingly stressed and could become a key factor limiting continued growth. *Incremental solutions such as building more AC and HVDC transmission can be done in specific high value locations but cannot easily be done at the scale that would be needed over the next decade to support the existing centralized grid operating model.* We have also seen that it may not be the most economical approach, especially as the cost of PV solar and wind continue to decline. A more distributed generation resource that is closer to the load centers may make more sense, providing a cost-effective non-wires alternative that can possibly also reduce the need for so much transmission.

The first objective for grid augmentation should be to maximize utilization of existing assets and infrastructure. New grid control devices, including power-flow control devices such as Smart Wires [6] and other Flexible AC Transmission or FACTS devices, could be deployed to augment existing bulk system infrastructure, allowing higher flexibility and controllability leading to improved system utilization, and the ability to continue to realize the core performance that the centralized grid provides, even as DER penetration increases. New technologies to dramatically increase the capacity of existing transmission lines and corridors, such as superconducting cables being developed by companies such as Veir, can be instrumental in getting more energy into cities and industrial load centers using existing rights of way.

Dynamic balancing provided by hydro plants (especially when paired with PV) and long duration energy storage from pumped hydro can provide strong grid support in some key locations where such assets exist. Hydrogen from renewables or using permanent carbon sequestration is likely to become available within the next decade and can provide baseline generation when DERs are not available, in some cases replacing existing natural gas with hydrogen [7]. Small modular reactors (SMR) can provide similar functionality, provided NIMBY (Not In My Backyard), cyber-physical security, and long-term nuclear waste management issues are addressed. Large utility scale PV farms and wind resources in conjunction with 4–8 h of energy storage can make them dispatchable and capable of managing key requirements such as ramp-rate control and loss of generation contingencies. Such hybrid plants can operate as a normal dispatchable generation resource but can also provide additional grid services such as spinning reserve, frequency regulation, grid-forming inertial support, peaker plants, and black-start capability. The grid operator will continue to manage the bulk system to ensure that operating costs are minimized, and reliability is maximized, even as DER penetration increases.

Such a system will see variability in resources with higher IBR penetration, which could impact system stability, and yield very low utilization for all the new elements that have been added. There will also be a dramatic increase in peak generation capacity that is connected to the grid, but with poor capacity factor (and we haven't even counted generation on the distribution side yet). Utilities can continue to operate their retired coal and gas generators as "synchronous condensers" and can retrofit existing large power transformers with fractionally rated inverters and energy storage to provide voltage and inertia support for the grid so as to improve grid stability—the transmission lines are already there. Transmission transformers can be retrofitted with power converters and storage to provide the grid forming and stabilization capability that the future grid will need. Similarly, the gigawatts of new PV and wind inverters being deployed can be integrated with energy storage to provide grid forming and inertial support capability, as well as can help improve grid stability. Some of these advanced functions are not available in existing inverters yet, but with some luck, these could be standard features within the next few years.

It should be noted that the above discussion looks at using existing grid resources to the extent possible, augmenting them with advanced control or generation/storage capabilities, all based on new technologies with steep learning curves. On a good sunny and windy day, things will be okay—but what about the other days, and what about when a "Hurricane Maria" or a wildfire impacts the bulk system, and it is not able to meet the load? What happens when the peak demand in an urban center is driven by a factor such as peak EV charging load and exceeds the level of supply that is available? Remember, we will not have operational coal and nuclear plants anymore. Under such conditions, we may have to look to the distribution system for support.

Uprating Distribution System Capability

The 2021 NASEM Report on "The Future of Electric Power in the United States" talks of a "new modular, distributed and edge-intelligent paradigm that competes with the existing grid" (Recommendation 5.2), and a system where "reliability and resiliency are achieved at the distribution level, while the bulk power system delivers low-cost energy" (Finding 5.4). It also goes on to say that "distributed generation (DG) at the grid edge represents a vast untapped DER for power balance and resiliency" (Finding 5.5). EPRI similarly talks of a transition from a DER agnostic to a DER-dependent distribution grid.

But as we have seen, this will not be easy.

The biggest challenges are—the integration of large amounts of non-utility-owned PV solar and energy storage with grid operations; the ability to manage and balance the peak demand posed for charging of all types of electrified transportation and other fast-growing new electrical loads with available generation capacity; and the need for resiliency when the bulk system is unable to deliver the needed power. In an unlikely but increasingly possible situation, a large area distribution grid may be cut off from the bulk system (the resiliency case) or may need to be islanded by the utility for balancing flagging generation with current demand. In such instances, the islanded system may need to operate for an extended time with local resources. The issues are similar for rural and urban distribution, but may be more acute for urban grids, because of high population density and challenges with significantly expanding the incoming transmission lines and the distribution grid.

Increase in rooftop solar and EV charging can stress the grid-edge, causing voltage volatility and overloading transformers and distribution lines. With poor or no visibility on residential distribution feeders, and no control over the peak load or returned power from a customer, most utilities are not well prepared for this transition. Extending the paradigm of local generation to commercial and industrial plants, we see that the problem will be multiplied. A warehouse with a 100,000 sq. ft. roof with PV solar and energy storage can generate around 2 MW of peak power and over 10 MWh/day, at a price that over the next few years will be at or below grid parity. If the generated energy is returned to the grid in an uncoordinated manner, this can pose a big challenge in terms of grid integration. On the other hand, if coordinated and controlled properly, this could be an asset. This local generation can also be used to charge 50 e-semis or 200 EVs per day, or to power 200 homes in an emergency! Many of these new grid-edge customers, such as datacenters, logistics centers, and large stores and shopping centers, are being built with integrated PV, energy storage, and local generation—echoing Finding 5.5 from the NASEM report. However, we do not as yet have the technology or regulations to allow seamless integration of these distributed resources with the grid, but this is very feasible over the next few years and can be a major factor by 2040.

To extend this concept a little further, even using existing warehouse rooftops for PV and storage can unlock tremendous potential. As an example, a city such as Los Angeles can realize over 3000–4000 MW of peak generation (per LA100 Report by National Renewable Energy Laboratory (NREL)) [7] from large commercial and warehouse rooftops. When paired with energy storage, this distributed resource can provide 30% of the city's peak demand and may also be able to meet peak EV charging demand, helping to balance the grid and provide resiliency at the edge—all without building new transmission or substations. The same strategy might work in other urban centers as well. But under the current utility operating paradigm and regulations in most states, this edge resource can only be used to provide a minimal level of grid services (and only for large plants, not residential solar), including frequency regulation and energy arbitrage—not microgrid functionality and resiliency services outside their own premises.

On the technology side, to achieve such a new integrated paradigm will require the next generation of grid-connected inverters with advanced grid-forming and collaborative controls, and the ability to build decentralized microgrids which can work in grid-connected mode under the grid-operator's or utility's direction as required; and can also operate as an autonomous decentralized microgrid serving multiple customers when islanded from the grid. On the regulatory side, for this to happen, regulators will also need to allow such microgrids to operate, and for utilities to be able to share some of the revenues generated. Today, most states only allow utilities to sell electric power, and microgrids cannot be operated where multiple owners/customers are involved. This has a big negative impact on the ability of a community or group of home/business owners to collectively achieve their goals for emissions, cost, and resiliency. It also requires a duplication of generation and energy storage resources by the utility to meet peak demand and achieve resiliency with low capacity-factor resources, dramatically increasing the cost of providing reliable service. Further, restricting "behind-the-meter" generation resources from supporting grid requirements also makes their internal business case very challenging. The ability to earn revenues from grid services can dramatically change the economic viability for these distributed resources—creating a win-win scenario.

Another major concern we have discussed earlier is EV charging infrastructure, which has to precede broad deployment of EVs. This may be feasible in urban areas with high population density and a large number of EVs, particularly if local generation (as seen above) is co-opted into providing this service. In rural areas, with large swaths of land, low population density, and few EVs in the early years, the establishment and maintenance of a ubiquitous charging infrastructure becomes very expensive. A solution that provides multiple value streams—e.g., grid support/resiliency along with EV charging—may provide better returns on the investment, attracting private capital for faster deployment. *A top-down program to simply install a lot of chargers, as seems to be the plan now, will probably succeed in installing the devices but may have a hard time showing economic viability and ability to scale.*

The distribution grid is clearly an area that has been neglected by the utilities for a long time and is in need of significant investment and upgrades. This includes hardware, intelligent sensors and reclosers, local energy storage, edge generation, including grid-forming functionality, and an ability to integrate millions of new edge devices that are intelligent and dynamic and can play a role in grid balancing and resiliency. With millions of intelligent cloud-connected devices, especially inverters that are manufactured in diverse countries with widely different technologies, and where each inverter can respond autonomously in microseconds, ensuring continued and reliable grid operation under a cyberattack also becomes a challenge. Traditional cybersecurity measures and processes that the utilities use for training, verification, and validation are important, but not sufficient for real-time-must-run (RTMR) situations in highly coupled systems such as a distributed and decentralized grid. We need to develop and validate cyber-secure control techniques for highly distributed and decentralized grids.

Looking at the economic and regulatory aspects, we see that as self-generation increases, utility revenues decline—creating a challenging situation for new investments that need to be made. If not tackled proactively, this can lead to the scenario where customer defection can cause a collapse of the grid system, resulting in chaos and huge impact on the most economically disadvantaged communities.

The role of the distribution grid operator is also clearly changing—from an entity that is the monopoly delivering electric power to all customers, to one that is the manager and coordinator of the grid (possibly including microgrids and customer-owned resources), where utility revenues can come from energy sales, but equally from the services provided to balance and stabilize the system, and to keep the system operational. A revenue model primarily based on energy sales is no longer sufficient. It should also be noted that a pure market-based model for a Disco (distribution company), where customers are free to choose an energy supplier, and the Disco's only role is to maintain the wires, does not work well, because there are no incentives to make any investments. We have seen challenges in places like Texas where such models have been used, where energy prices spiked under abnormal conditions, because no investments were made to ensure reliability and resiliency or to accommodate dynamics of new resources on the grid—that was not part of the compensation metric.

Energy Access and Greenfield Grids

Much of what happens in developed economies will require a compromise in reliable grid operations as changes are made to get to an emission free energy ecosystem, at the lowest cost and highest speed, though hopefully with minimal disruptions for customers along the way. *For greenfield systems in emerging markets and in countries where a majority of the people still live off-grid, a leapfrog strategy that is largely based on new exponential technologies is possible and should be seriously considered.*

As we have seen, energy access is not simply a problem of delivering energy, but of addressing issues of poverty, education, and livelihoods. Viable solutions need to be able to:

- start small and scale as demand grows, matching system capacity with ability to pay;
- operate across a wide range of demand-and-supply scenarios, including in grid-connected or grid-independent modes;
- deploy flexible plug-and-play modular solutions that are interoperable across vendors and technology generations;
- be maintained with minimal technical expertise in the field;
- support and grow the local economy;
- create no long-term issues in terms of waste management.

Providing energy access to all who need it is now possible with twenty-first-century exponential technologies at a cost point that is potentially much lower than a conventional grid built using twentieth-century technologies.

Providing energy access to all who need it is now possible with twenty-first century exponential technologies, at a cost point that is potentially much lower than a conventional grid built using twentieth century technologies.

Example of an Energy Access Bottom-Up Grid

By way of example, we have been working (along with many others who may be working on similar concepts), on the design of a modular scalable "Plug-n-Play-Grid" building block that converts a solar panel into a dispatchable PV microgenerator, able to supply standard AC power for a full day, even after the sun has set [8]. A single panel rated at, say, 400 watts could generate 2 kWh of energy on a normal sunny day—sufficient to provide Tier 2 service for eight to ten homes.

We have seen in Chap. 2 that even a modest "load diversity factor" can have a big impact on the peak rating of the generation system needed. As the demand grows or more homes are added to the service, additional panels can be connected together in an ad hoc manner to increase capacity as needed. Panels on different homes or businesses can be interconnected at will to create a flexible and fractal microgrid. The system can work autonomously and in a completely decentralized manner, balancing available energy/power with demand in real time, maximizing energy delivered from each panel, and implementing basic transactive functions such as billing and system optimization. At a community level, hundreds of the PV microgenerator panels could be connected together to power a village, scaling up to support commercial and manufacturing activities as needed.

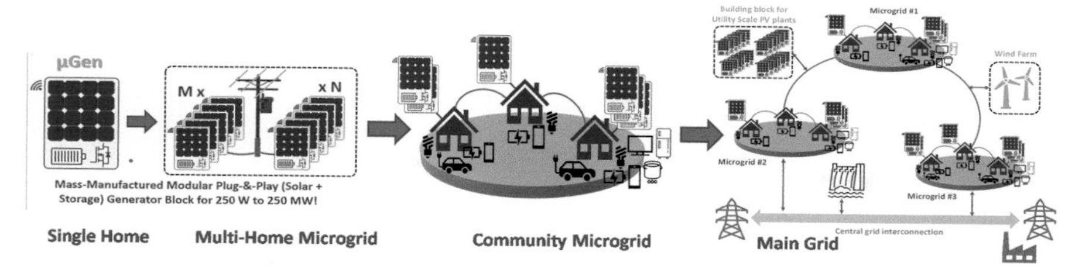

Fig. 8.1 Democratization of Energy—a modular, interoperable, plug-n-play building blocks to implement flexible ad hoc energy systems from single home to microgrid to bulk grid level

We envisage that such a framework based on Plug-n-Play building blocks and the right technologies, can also be used to scale up from a single home to multi-home to community scale to utility PV farm level at 10–100 MW and more, now including 4–8 h of energy storage. We would like that such a PV microgenerator, at a single panel or microgrid plant level, can be arbitrarily connected or disconnected to the AC grid when the grid arrives (only if such connection/disconnection is allowed by the grid rules). It can also be dispatched by the grid operator if needed. The fleet of PV microgenerators can also serve as a dispatchable resource for the bulk grid as and when required. Figure 8.1 shows a representation of how such a modular system could be built.

Such a PV microgenerator can only function as described above if key advanced functionality is built in. This includes the ability to operate without interacting with the grid or with other inverters; to autonomously operate in grid-connected or islanded mode; to provide grid forming capability and grid support as needed; to share real and reactive power dynamically with other inverters without low-latency communications; and to continue operating through major system faults and configuration changes.

Another part of such a finite-resource system that is needed in grid-islanded or microgrid mode is a "transactive" overlay, where each source and load is able to determine autonomously, in real time, whether the grid is facing a condition of resource surplus or scarcity, as reflected by a universal "price" signal, and to use that to determine the level of participation for the source and which loads need to continue to be supplied with energy. This dynamic balancing strategy, which can also be automated, can allow generation to follow load in time of surplus energy availability, and for loads to follow generation in time of scarcity, allowing continued operation across a wide range of operating conditions, including when the communications link fails or is compromised. And all this can be achieved with plug-n-play modular building blocks that have a high degree of interoperability and ease of installation and maintenance. The modular structure of the microgenerator also makes it easier for recycling at end-of-life. Such a system could be installed, operated, and maintained by a local entrepreneur or by a local "energy manager." And by starting small, the investment in the infrastructure could be coordinated with rising demand.

What the above discussion hopes to illustrate is that a fresh approach, built with twenty-first- century exponential technologies, can provide a new and cost-effective alternative to the traditional twentieth-century grid. Such a system can autonomously operate stand-alone as a microgrid and can also, without any retuning, operate while connected to the traditional grid, as needed.

While the World Bank and national governments are spending $16B/year on grid extension in the LDCs (a small fraction of what is needed), less than 1% of that amount is spent on solar home systems—even a step up to 10% earmarked as incentives would make a huge difference over a period of ten years in driving these solutions down the steep learning curve that exists. For instance, following the example of the Indian government's UJALA program, procurement targets could be set for say 20 million PV microgenerators (including storage), installed and serving customers at a blended rate of, say, $0.20/kWHr on the total energy delivered, with critical Tier 1 access guaranteed with a monthly connection charge, and Tier 2 and above at variable rates based on availability of solar power. The variable rates, easily correlated with available solar energy, can also provide an incentive for the use of lower energy consumption appliances.

Such an approach seems counter to the centralized grid extension policy that energy access generally takes today—a strategy that is failing to recover costs. Here we are showing what happens when you buy into the argument that solar energy will be abundant and very low cost, almost free at some times, making it an ideal building block for scaling as needed, without the massive upfront investment on day one for a national scale central grid that reaches into every village and community. Clearly, this approach will not give four nines reliability levels in the early years, but then, compared to off-grid, this is still much better.

Critical Tier 1 functions such as lighting and digital connectivity can see much higher reliability levels of >90%. As commercial and manufacturing customers join the customer group, they can put in local generation, including expensive and CO_2-emitting diesel generation as the ultimate backup source if absolutely needed. At the top-down central grid level, additional zero-emission resources, including wind, hydro, geothermal, hydrogen, and others can be integrated as needed to meet overall system objectives. It is interesting to note that this approach also provides a blueprint for how a future decentralized grid could be built in developed nations as well.

Getting to Decarbonization Needs More Than Technology

In this chapter we have begun to see viable pathways that can get us to a utopian, rather than dystopian, energy scenario by 2040, which can then help us, through market forces, to hopefully get to the desired 100% decarbonization goal that has been set for 2050. That is a point in time by which we can significantly impact how society generates and uses clean and sustainable energy, while at the same time addressing issues of economics and decarbonization. We have shown that while exponential technologies, supported by hundreds of billions of dollars in available investments, can move and grow fast, the grid which is the glue that connects everything together in this future decarbonized world will be highly challenged to evolve at the required pace.

If utilities and regulators do not move quickly and steer the architecture and evolution of a new future grid that has the characteristics we need, it is likely that the exponential technologies will grow in an uncoordinated manner, and we will then have a scenario that is chaotic and expensive. A preferred solution is to tweak the regulations slightly so that regulated utilities and the technology suppliers can work together, using existing grid assets to the extent possible, and augmenting them with additional layers of twenty-first-century clean technologies, so as to ensure a smooth transition to a desirable future energy system. As we have discussed, this should be driven by aligning both our economic and climate change objectives—for the first time ever, the two are not in direct conflict with each other.

For the first time ever, economics and decarbonization are not in direct conflict with each other.

It is highly optimistic, and perhaps not realistic, to think that we will be at net zero carbon emissions by a specific date (certainly not 2040), but if the alignment and incentives are correct, the system will automatically move at the best speed possible (based on technology, scaling, economics, and policy) toward broad holistic societal goals that are set and measured/tracked to ensure they are achieved. Choice of objectives that are too narrowly specified can generate a lot of unanticipated consequences and may trigger the tragedy of the horizon—it would be better to avoid that by design. We want to caution against techno-optimism, where we assume that the simple existence of technical capability solves all problems instantly—but we also do not support techno-pessimism either, where one feels that change at the scale we want is simply not possible, just because it is a heavy lift, and because we have not accomplished it in the past.

We want to caution against techno-optimism, where we assume that the simple existence of technical capability solves all problems instantly—but we also do not support techno-pessimism either, where one feels that change at the scale we want is simply not possible, just because it is a heavy lift, and because we have not accomplished it in the past.

We have seen that the last 20 years have shown us and reshaped our thinking on how rapid change at scale can occur simultaneously in many different areas. We have discussed at great length the fact that exponential technologies have defied and continue to defy predictions in terms of how they can grow. PV solar production, starting at <20 GW/year in 2010, grew to 400 GW/year in 2023, and could reach 1940 GW/year by 2030, if current growth rates are maintained (and there is no reason to believe they will not be—in fact, they could be further accelerated, for instance, to make hydrogen at scale). At the same time, the learning rate of 24% suggests that the price could be reduced to as low as 25% of 2020 LCOE levels, which are already significantly below grid parity today, even when storage is included!

It is simply amazing to note that the total global installed capacity of solar could significantly exceed 8000 GW by 2030, 8× the total US generation capacity today. *This massive change on a global scale is occurring at an unprecedented pace that no one was able to predict—not even the techno-optimists!*

Government research funding to national labs, as well as universities and industrial research centers, is a critical enabler of the thousands of new scientific discoveries and technologies that have the potential to completely transform our world, all driven by an exponentially expanding base of scientific knowledge and capability. US companies were once the envy of the world, and the reason that the US won the Cold War and the space race—with innovation driven by federal funding, and with massive disruptions, such as nuclear power and mobile communications. But, once US companies lost their long-term R&D capability (with the possible exception of the military-industrial complex), the process for translating science to technology to market for truly disruptive applications in "hard science" areas was significantly impaired. Two notable exceptions are the military-industrial complex and fields related to digitalization and the biomedical field (which we have already discussed).

As universities and national labs began to pick up the mantle of fundamental research, the need was identified for a new T2M process and viable commercialization models. Several states launched their own Departments of Economic Development, but these were still aimed at simply bringing new business to their states, not to nurture new disruptive technologies.

The only strategy we have had for commercializing deep-tech was a random process where we waited for a happy intersection of a capable entrepreneur and an angel investor, who in turn bumped into the underlying science at a university research lab or national lab. As we have seen with dozens of start-ups that were based on technology from a university or national lab, the road is rocky, and the process can be very inefficient. The success rate for such deep-domain disruptive start-ups is extremely poor, much lower than the already very low rates for normal VC-funded startups.

In Chap. 7, we have proposed that there needs to be a more structured and intentional process for this translation, one that is centered around the research centers where the deep domain knowledge resides, and who are able to use the funding in an efficient (frugal) way and are able to significantly accelerate the de-risking of the underlying technology and business models.

The key objective in the early years of doing research (under an assumption that the underlying science will translate into commercial success) is to make reduction of innovation and technology risk a key objective and outcome of the core research efforts (which it is not today). This can be achieved through a holistic storyboarding process, as mentioned above, that addresses key questions regarding potential applications, important specifications and requirements, competitive advantage, how scaling will occur, whether learning curves are likely to be steep and sustainable, how value can be derived, and possible business models (e.g., incremental or disruptive)—and does so at an early stage of the project when the research center is still deeply involved, and before any commercialization commitments or investments are made. For maximum impact, it is likely that the scope of what can be done at universities or national labs may need to be redefined, to allow risk-reduction as a normal conflict-free activity.

Further, as discussed in Chap. 7, there is also a need for a "bridge" organization that sits between the research entity and the real world, which enables field testing and customer feedback, as the final stages in validating the value streams, and helps in streamlining the translation process and helps to accelerate scaling of the disruptive science that was developed and demonstrated. In fact, addressing these questions early can also guide the research itself to make sure that the questions the research is trying to answer are correctly framed. This is in no way trying to take away from scientific rigor or academic independence—but rather to make sure that when public monies are spent, the potential for possible public good is a valuable end result that deserves exploration.

It is pertinent to note that there is significant downside risk to "business as usual." Left to themselves (and once they are committed to exploiting a new technology), industry typically seize the IP they have licensed or developed, as well as the technology competency that they possess internally, and pursue a strategy for rapid market penetration and dominance, one that minimizes their costs and internal disruption, and maximizes their profits. Specifically, in the large markets, competition can be fierce, but only after the problem has been shown to be solvable and innovation risk is reduced (investors and their companies often want to be at the "leading" edge, not the "bleeding" edge—so many of them like to be "fast-followers"). Such a strategy automatically generates incompatible proprietary solutions, causing confusion in the market and slowing down the rate of scaling. Over time, this competition plays out, with some winning and some losing, and a gradual steady state position is reached. As we have seen, such a market-driven process is very inefficient and can take time, time we do not have.

Good policy and incentives can have a big impact on accelerating the scaling of economically viable solutions that we need today to address issues of economic growth and climate change. Policy should favor solutions that ride on steep learning rates and are improving rapidly in performance, almost on a year-over-year basis, but should also recognize that such an approach would create uncertainty in outcomes. For the best results, objectives should be technology agnostic and based on a high-level articulation of value delivered and metrics to measure it, and not on a narrow technical specification. As we have discussed, the velocity of change, sustained learning rates, and uncertainty also lead us toward solutions that are flexible, modular, distributed, and interoperable, and deliver on the value that the policy targets. Interoperability, especially for physically linked devices such as grid-connected inverters, can be challenging between solutions from different vendors, but becomes critical for rapid scaling.

As an example, all EVs should be able to use the same charging infrastructure and all IBRs must support grid needs and be interoperable with each other when they are grid-connected, if they are to benefit from government funds. Similarly, low dependency on critical raw materials and end-of-life management become important elements that only policy can drive. Finally, policy also creates incentives (preferred) to help achieve key targets (such as resiliency and equitable distribution of benefits of the energy transition) and penalties for people who violate the public good (e.g., long-term public liability due to nuclear waste management or health impacts due to poor waste management). As in the German FIT program, incentives can start strong and be reduced over time as parity with existing non-compliant solutions is achieved.

Similarly, policy can also incentivize, subsidize, and penalize many of the wrong things, with poor correlation to positive societal impact. This is a reflection of the power and politics of incumbents and the policies from which they personally benefit, and which they will not let change. Examples abound, including coal, oil, pollution, nuclear—the list goes on! Policies can accelerate, and can retard, the adoption and scaling of the new technologies that can save us. They have a huge role to play.

Bioethanol for Transportation

An interesting eye-opening example relates to the use of bioethanol as a fuel for transportation in the US.

- In the US, bioethanol is used as a fuel for transportation, with heavy subsidies encouraging farmers to grow corn (or other cellulosic feedstock). This requires 21 million acres of land, consuming as much as 600,000 gallons per acre per year of water, which totals to 12 trillion gallons (3400 times the water stored in a full Lake Mead), water that could be used for human consumption or to grow food crops. Remember, bioethanol is also a very low efficiency solar to fuel energy converter.
- On a per acre basis, PV panels can generate sufficient energy to power more than 300× the number of EVs, as compared with bioethanol used to power internal combustion engine cars (this is due to a cascading of inefficiencies related to solar-to-phytomass conversion and the Carnot cycle) [9].
- Farmers could also potentially earn much more from putting PV panels on their land than growing corn for bioethanol and could still have land left over to grow food for human consumption. If the land used for corn/bioethanol was converted to PV solar farms, 17,000 GW of peak power (17× of US peak generation today) and 25 TWHrs (6× of total electrical energy used per year) could be generated! There seems to be no defensible reason to maintain subsidies for bioethanol—but it persists, and to the tune of $116B over the last ten years!

Conclusions

With such a wide canvas, so many players and so much at stake, what can we do? Our problem seems to be too complex and wide ranging to be solvable—which, to some degree, is true. This problem cannot be solved with guaranteed outcomes.

However, if we can all agree on the aspirational goal of where we would like to be—what we have described as our North Star—for directionality, and we take a broad and holistic view of issues as we take actions within our individual silos, anticipating challenges and correcting our course as new inputs are received, then maybe we can accomplish the best and desired outcomes.

If instead, we continue to operate within our own silos per our static long-term plans, ignoring adjacencies and working within a narrow and self-serving world view, then we will likely have to live with a dystopian world that we ourselves will have created and be responsible for. *For the first time in our history, we can think about meeting the needs of all our families in terms of economic growth, even as we work to decarbonize and combat climate change.*

For the first time in our history, we can think about meeting the needs of all our families in terms of economic growth, even as we work to decarbonize and combat climate change.

A few of the abstracted key points discussed in this chapter are summarized below:
New science and technology-based knowledge and discoveries will keep growing, driven by scientific curiosity and global research that cannot be stopped, and which can in some cases be the basis for new exponential technologies that follow steep and sustained learning curves.
Some of these exponential technologies provide a unique "once in a lifetime" opportunity to accelerate the energy transition, using the economic opportunities created by expanding markets and rapidly declining costs to drive a virtuous cycle. However, the interactions of these exponential technologies with adjacent sectors can create roadblocks to continued sustainable scaling and growth and create unexpected and/or unwanted outcomes if not managed properly.
These new exponential technologies can dramatically expand our viable options, fundamentally altering our dependence on geopolitically sensitive natural resources, and moving us to an era of abundant, economical, equitable, safe, and sustainable energy.

There is an opportunity to direct and accelerate some of these developments so that we can expand the palette of scalable solutions to include those that are both economically viable and sustainable and can help in our decarbonization initiatives.

There is a need for new processes and mechanisms to rapidly assess the scaling potential and impact of promising exponential technologies that are emerging from research universities and national laboratories, and to do this on dramatically reduced timescales to achieve significant de-risking, particularly in terms of their potential to ride on steep and sustained learning curves, and to reach scale.

As technologies evolve fast and are adopted and scale over short timelines, there is a need for a workforce that is trained in the new technologies and is also open to life-long learning.

Similarly, there is need for a new generation of entrepreneurs, industry leaders, and investors who understand risk mitigation and can manage companies that are driving disruption based on deep-domain science and technology innovation.

Policies should encourage technology-agnostic value-based outcomes, using metrics that reflect a holistic view of the issues, and are the basis for temporary incentives that can accelerate scaling of the right type of solutions to stimulate overall economic growth and accelerate the time to parity with solutions that are less desirable at a societal level, while still ensuring safety, sustainability, equity, reliability, and resiliency of the new energy system. Of particular interest are solutions that are modular, interoperable across vendors and technology generations, have well defined life-cycle costs and rely on abundant materials, which do not require massive inflexible infrastructure to operate, and which follow steep and sustained learning curves.

Most of the technologies being developed are "stand-alone" and can move at high velocity. However, the electricity grid is the hard-wired physics-constrained network that connects most energy resources and loads. It is a highly regulated, and at the same time fragmented, structure that has been designed and constrained to only allow it to move slowly.

How the grid evolves over the next 20 years will have a significant impact on the future energy infrastructure, and our ability to meet economic and decarbonization objectives. If the approach taken is to expand the current grid under today's operating paradigm and fragmented regulatory structure, we are likely to see many challenges in the years ahead.

The best outcome may be to build on the existing grid, augmenting the wires, substations, and carbon-free generation that already exist with new exponential technologies, while ensuring that the resulting system is flexible, dynamically controllable, and able to meet fast-changing needs. This future "*grid as an ecosystem*" will require a host of new technologies and capabilities but may minimize the build of new bulk generation and T&D infrastructure, a process that is expensive and challenging and takes a lot of time.

There is sufficient motivation and funding to achieve these goals today. There needs to be clarity and flexibility and an assurance that all players are aligned in the end goals and objectives we are pursuing. It is interesting to note that following economic principles, accelerating the translation of exponential technologies with steep learning curves, and creating a set of lightweight incentives, guardrails, and policies can set us on a trajectory to sustainably solve our climate issues, even as we move toward energy abundance, affordability, and sustainability.

References

1. Lambert, F. (2022, August 18). *Tesla's virtual power plant had its first event helping the grid—looks like the future*. Electrek, and Sila Nanotechnologies, Quantumscape.
2. Rauwerdink, A. (2019, July 23). *Heavy industry decarbonization*. National Academy of Sciences.

3. UNIFI Rules – Divan.
4. Grid as an ecosystem – Divan – DOE Presentation, July 2023.
5. Breakthrough Energy Ventures. https://en.wikipedia.org/wiki/Breakthrough_Energy
6. Smart Wires, Technologies for a Net Zero Grid. https://www.smartwires.com/
7. LADWP, 100% Renewable Energy Plan Study: http://www.la100study.com/
8. UN Global Solutions Summit – 2023, Divan.
9. Energy 2030 Conference Organized by Georgia Tech, and IEEE Spectrum - Divan.

Beyond 2040: Getting to Energy Utopia

9

Getting to Energy 2040 (with Desired Outcomes)

We are rapidly approaching a unique point in human history, possibly a singularity, where our lives will potentially be irreversibly transformed by a continued, accelerating, and unstoppable explosion of new scientific discoveries and technologies. The processes and institutions that brought us up to this point are seemingly challenged to cope with the velocity and ubiquity of this change. Even the human brain is unable to manage the massive flow of often conflicting information that it needs to process, sometimes in real time, leaving us with uncertainty about what should be done and what the final outcomes will be.

But, at the same time, everything we have achieved as humans is deeply intertwined with our ability to access energy when and where we need it. Human history from prehistoric times, certainly for the past 6000 years, has been built on the foundation of energy, and our ability to access and control it. It fills our needs, gives us comfort, allows us to do what we want, and is often the basis for projecting political power.

Energy has also been a story of haves and have-nots, of inequity that was exacerbated by the geopolitics of energy resource availability, and of economic development that was directly tied to the access to adequate levels of affordable energy. Entire continents have been left in the dark and economically underdeveloped, a condition that is difficult to cure. There is no doubt that energy use is inextricably linked to human progress and to continued improvements in terms of a human development index (HDI), and access to energy should be a fundamental right for all people.

The industrial era's pursuit of energy based on fossil fuels, the only compact, portable, and dispatchable source of energy available until recently, has resulted in massive levels of anthropogenic carbon emissions, which have in turn been linked to global warming and potentially cataclysmic climate change. So, even as our innovations have made our own lives better, we may have made life for our children and

D. Divan, S. Sharma, *ENERGY 2040*,
https://doi.org/10.1007/978-3-031-49417-8_9

grandchildren much worse. This linkage has become more visible and better understood over the past few decades, especially during the last 20 years.

In an attempt to reduce the level of carbon emissions, the world has tried to unite behind concerted action plans, developed over a series of international agreements, in the hope of holding temperature increases to within 2 °C. Such a temperature increase is considered by most climate scientists to be critical, beyond which the chance of runaway global warming becomes very high. Setting goals for addressing climate change that are based in abstract science (e.g., reducing carbon emissions) often do not relate well to people's daily lives, and have made collective action almost impossible, because there have been no viable pathways or options that everyone could rally around.

Setting goals for addressing climate change that are based in abstract science (e.g., reducing carbon emissions) often do not relate well to people's daily lives, and have made collective action almost impossible, because there have been no viable pathways or options that everyone could rally around.

As we have observed before, no one wants to deliberately generate CO_2, it just happens to be an undesired by-product of people living their daily lives. We suspect people would happily adopt energy options that did not generate CO_2, provided they could get all the energy they wanted, and the energy was also cost-effective.

Strident and urgent calls for limiting carbon emissions, accompanied by an inevitable doomsday scenario if emissions are not controlled, have only generated a sense of helplessness, and in some cases, leading to apathy. Until now, any solutions that could slow down carbon emissions required either a dramatic reduction in economic activity and standard of living, or a dramatic increase in the cost of solutions, if they were non-carbon emitting. This has proven impossible to implement at the scale that is needed.

A vast majority of people were not prepared to suffer near- to midterm economic loss and hardship, simply so that a potential future problem could be avoided—the Tragedy of the Horizon. As the book *Speed and Scale* by legendary investor and VC John Doerr indicates, "*Fundamental changes do not happen because they are virtuous. They happen because they make economic sense. We've got to make the right outcome the profitable outcome, and therefore the likely outcome*" [1].

This right outcome, from an environment and climate standpoint, has not been possible with twentieth-century technologies. However, for the first time ever, we see an opportunity to create a win-win scenario today. We can align our economic goals with decarbonization by creating solutions that are both "wallet-friendly and planet-friendly" [2].

This dramatic shift in capability rests on a deeper understanding of the science and technology of new materials at the atomic level, and our ability to manipulate and manufacture microminiaturized devices and components at scale—capability which has matured over the last couple of decades, and which continues to develop at an ever-accelerating pace. This in turn has brought us the technologies we need, technologies with steep and sustained learning curves that have resulted in an exponential growth in knowledge. These factors, taken together, have fundamentally changed what we as humans can achieve.

We have also seen that change in the energy ecosystem, especially the pace of change, is driven by a complex interplay between science, innovation, technology, entrepreneur-disruptors, industry, finance, policies, politics, geopolitics, and of course the billions of energy customers who need to accept and embrace the change being foisted on them. On the other hand, change is resisted by those who stand to lose, including the incumbents who hold vast financial and political power, and who will resist change with every tool in their toolkit. How long the transformation will take and how smooth (or chaotic) the transition will be is difficult to predict, but historical precedence suggests that it could take 50+ years if we followed past pathways. This is time we simply do not have. We must do better.

Given the complexity of the energy ecosystem and the diversity of players involved, we have, in this book, tried to refrain from oversimplifying the issues and have spent considerable time on some of the nuanced factors that may be critical for success, especially in terms of electrification. We have also endeavored not to unduly complexify the problem and have extracted and prioritized overarching themes wherever possible.

We are also reluctant to overspecify what the solution should be, keeping in mind that there is a broad slate of solutions becoming available, and there may indeed be many different ways to solve the broader problem. With so many fast-moving technologies in so many adjacent areas, we have seen that it becomes impossible to predict with any certainty, how the future will actually evolve, even a short 20 years out!

The approach we favor is to create a clear and aspirational vision of where we want to go, in humanly relatable and technology-agnostic terms. Such a "North Star" can provide directionality, guide our decisions and actions in this rapidly evolving world, and resolve confusion and ambiguities as we take action within our own silos, and in our day-to-day lives.

A good starting point is to define a top-down vision for what we want to achieve in the field of energy, which can then hopefully lead to a roadmap of how to get there.

We suggest the following vision:

"A global economy that is fueled by abundant sustainable energy with no greenhouse gas (GHG) emissions which powers an energy ecosystem where the energy is also Flexible, Resilient, Economical, and Equitable (FREE!)."

VISION: *"A global economy that is fueled by abundant sustainable energy with no greenhouse gas emissions which powers an energy ecosystem where the energy is also Flexible, Resilient, Economical, and Equitable (FREE!)."*

Such an energy ecosystem must meet the needs of the fast-evolving sources and loads that it serves, while also ensuring that the rights of future generations are preserved. Another critical component of such a broad vision is value-based metrics that allow us, in technology-agnostic human terms, to assess progress and success.

At first glance, such a vision may feel overly ambitious and at too high a level to meaningfully guide our day-to-day actions in the trenches while we are designing EVs, solar cells, articulating policies, and business models, or managing companies and investment funds. However, we believe that such a shared holistic vision helps to create alignment between communities which, while intrinsically very different, all intersect with each other, and are all critical to achieving successful outcomes at scale, especially if we desire an accelerated timeline.

What is important to note is that unlike in the past, today, we have the distinct possibility that this vision can actually be achieved with technologies and solutions that we either already have, or that are already visible in our research labs, and could be available at scale in the very near future. Many of these are exponential technologies with steep and sustained learning curves that meet many of the criteria stated above, and which are already at economic breakeven, or are showing very viable pathways to getting there, and to achieving rapid scaling. As this unique moment in time is better understood by VCs and investors, the flood gates of investment capital are turning on, with over $500 billion in venture capital and $3.2 trillion in private equity funds committed to funding the upcoming disruptions [3].

This big first tranche of funding is aimed at supporting the commercialization and scaling of new scientific discoveries and technologies that solve critical challenges, have the right attributes, and have been substantially de-risked from an innovation, technology, manufacturing, and business model perspective. However, we have also seen that with the departure of many large energy technology companies from the arena of long-term high-risk disruptive R&D, the translation process from science to impact is broken and needs to be fixed.

As we have discussed in this book, this may also be an opportunity for major research universities as well as national and industrial research labs to redefine their role in the innovation ecosystem, making sure that their work also includes solving the big problems facing society today, that they develop the skills internally to demonstrate and to accelerate the de-risking and translation of these technologies, and that success is measured by meaningful impact at scale, and not by narrow metrics such as a high number of incremental peer-reviewed publications in high-impact-factor journals, or an arbitrarily defined technology readiness level (TRL). Such an approach needs to be supported by governmental and institutional R&D funding, which lays down criteria that are aligned with the grand vision and measures success and potential impact with value-based metrics.

As we have also seen, a likely scenario for how this decarbonization can be done is based on massive electrification of many sectors, using renewable energy, green hydrogen, or another emission free generation resource, as well as the replacement of industrial heat sources with green hydrogen or e-fuels made using carbon capture and utilization (CCU) principles. This provides a viable approach to convert our primary energy resource, from what has mainly been GHG emitting fossil fuels, to energy that is not only dispatchable and free of carbon emissions, but also sees continuously declining prices. Taken together, it is clear that these drivers can completely disrupt the existing energy ecosystem.

The economic drivers are so strong that we believe this transformation cannot be stopped. It is also clear that this will represent a major disruption for the incumbents, who have invested heavily in twentieth-century technologies—but this is just a normal Schumpeterian disruption at work. The challenge is whether we can achieve this with minimal disruption to the daily lives of the broader population, or whether this will involve a chaotic transition period where energy is more expensive, less reliable, and emits even more CO_2 than we do now.

We have seen that the electricity industry has generally evolved, driven by regulations and policy, into a risk and innovation averse, slow-moving sector that is challenged to manage the massive fast-moving change that is being thrust on it. A continued move toward a future grid that is centralized and operates with today's paradigm is likely to see major challenges, as it intersects with new technologies that are more distributed, decentralized, and dynamic. Such an approach will also neglect the massive investments in clean generation and storage that are already taking place behind the meter, resulting in massive duplication of resources and making the transition very expensive.

A better approach may be to augment the existing grid infrastructure with new dynamic grid control devices and a new generation of inverter-based resources (both in front of and behind the meter), that autonomously act in a coordinated manner (with light supervision from the grid operator), can be rapidly deployed, can improve utilization of existing grid assets, and can allow the evolution of a new grid as an ecosystem. This needs a uniform set of forward-leaning policies and incentives, with appropriate guard-rails and monitoring for compliance, which ensure that the overall vision is not compromised. There needs to be an acceptance that there are likely to be reliability and integration issues during the transition period, but to counter that with a strong focus on rapidly rectifying the problems so that overall performance and reliability continuously improves. Such an approach is common in the commercial world but can create challenges in the regulated world of electric utilities.

On the other hand, if the electricity industry cannot respond in a timely and effective manner, disruption will still happen—only now it will be the grid that we all rely on that may be disrupted. And there may be no quick fix for that problem! If this results in a collapse of the current electricity system (just as mobile phones did to land-based phones), the trillions of dollars already spent on today's grid infrastructure will need to be written off. This can make the energy transition more expensive than it needs to be (it should be noted that much of the existing electricity infrastructure is old and has already been amortized and written off and will not impact computation of returns on the new investments being made by the disruptors). More importantly, such a pathway will take away precious time, time that we do not have.

We also feel that new emerging economies, especially the least developed countries (LDCs), that have little by way of electricity infrastructure today, can be the first to leapfrog our twentieth-century grid, and to demonstrate how we can build a new sustainable energy ecosystem where a basic level of energy access is "FREE"! In such a system, higher reliability levels will cost more, depending on the level of reliability needed, but could still be lower cost than we have today. What would such a system look like and what are its characteristics? Can we conceptualize such an energy utopia, as well as the path we must take to get there? We are not trying to look into a crystal ball to predict the future, but to describe what one such desirable future could be. Hopefully we can get there in the days after 2040, provided we act wisely and quickly in the days that are just ahead of us.

Accelerating the Journey to Decarbonization

We are on the cusp of the next energy transformation, where we can rapidly transition from a fossil fueled society that has struggled to obtain and equitably distribute energy to all its people, to one that has abundant, clean, equitable, and affordable energy for everyone, energy that comforts and sustains life, and does not pose long-term hazards and remediation costs for future generations. Done right, this can also set us on the path to decarbonization and give us the technology solutions that we need to adapt to and eventually mitigate the impacts of climate change. Hopefully we are learning from our past mistakes and will ensure that forward-leaning policies and incentives provide the guardrails for scaling the new innovative solutions that will come out of our research laboratories and institutions, and which will be commercialized by entrepreneurs, investors, and companies.

With the fast pace of change that we are seeing today, massive investments in huge infrastructure that take years to plan and build, and which are outdated and too expensive by the time they are built, is probably not the answer. What is needed are solutions that are modular, distributed, flexible, and rapidly deployable, and which results in an energy system that is resilient and can adapt and interoperate with the new technologies and changes that will inevitably follow. We need to avoid solutions that create long-term problems for future generations to manage, and that disturb the natural energy equilibrium of the planet. We have already seen the impact of burning fossil fuels. Nuclear waste creates a 10,000-year problem. Carbon capture and sequestration (CCS) can, if not done properly, leave behind caverns full of gigatons of CO_2 that can last for thousands of years and potentially leak out due to accident, natural disaster, or malevolent act.

Similarly, preference should be given for solutions that are intrinsically safe and do not cause extreme hazard for a community or large population base, when and if accidents or "incidents" do occur, even if those are infrequent or rare. Think thousands of tons of hydrogen stored in salt caverns, nuclear reactors damaged by a tsunami or system failure, microreactors in insecure places where malevolent actors can gain access to radioactive materials, ammonia that can leak out of a tank, a natural gas pipeline that is breached, or a deep-sea oil well that is damaged and leaks oil into the sea. The same applies to large dense battery systems located in densely populated urban areas, where the batteries can catch fire, a fire that is not easily put out.

We also need to holistically consider cradle-to-grave lifecycle costs, including recycling, waste management, land restoration and tailings management for mining operations, methane and hydrogen leakage and flaring, remediation of water and grounds impacted by improper waste management, and potential health impacts for communities. Batteries and solar cell recycling also need to be on top of the list. When looking at technologies such as biofuels, we should also consider environmental impact, including water, land, and fertilizer use, and whether the same function could have been done in a more efficacious manner. We clearly cannot solve all the problems in one stroke, but measuring and monitoring the various offending elements, putting a remediation price on it, and using the funds collected to provide incentives for solutions that can ameliorate the problem, can move us toward long-term sustainability.

> *We clearly cannot solve all the problems in one stroke, but measuring and monitoring the various offending elements, putting a remediation price on them, and using the funds collected to provide incentives for solutions that can ameliorate the problem, can move us toward long-term sustainability.*

The backbone of a future energy system (at least until cost-effective and demonstrably safe fusion energy becomes viable) should be based on plentiful very-low-cost sun and wind energy—the new "free" primary energy resource. This can be augmented by other clean energy resources, such as energy storage, hydropower, and green hydrogen, to provide the level of reliability that is desired, and that we have today from our fossil-fuel-based energy ecosystem. The apparent duplication of resources suggests that if the transition is not done right and the future energy ecosystem simply mirrors today's system, we will have a system that could be higher cost than our current system. But that does not have to be the case.

While we have discussed many new trends and technologies in this book, we wanted to highlight a few that we feel can have a significant impact and can potentially redefine the energy ecosystem of the future. All these trends assume that the current grid can be upgraded to manage these new resources and loads—which, as we have seen, is not all that certain! Some of these new emerging trends include the following:

- If the current learning rate of 24% for PV solar is sustained until 2040 (as seems likely), we can expect to see LCOEs of <$5/MWh. When the sun is shining, the next kWh of energy, after energy-storage reserves are replenished, can be virtually free.
- Similarly, energy storage costs are plummeting, expected to reach <$80/kWh, and possibly much lower, over the next few years. PV and wind integrated with storage will become the norm (as is already happening).
- Hydropower today provides ~100 GW of existing generation capacity in the US, coupled with many weeks of energy storage in the water behind the dam. Pairing

hydropower with local PV creates a low-cost dispatchable generation resource, with integrated long duration energy storage, but where the hydropower is only used under emergency conditions. This is doubly beneficial as the hydro plant is already connected to transmission and is under control of grid operator dispatch. In some cases, it may also be possible to convert the hydro plant to pumped hydro.

- Transportation, we feel, will be electrified, and will not generate net carbon emissions. Many possible solutions exist today, including batteries, hydrogen, and carbon-negative fuels. Although electric aviation is generating strong interest, it is possible that in the midterm, we will still need to fly with hydrocarbon fuels, especially for large aircraft flying long distances.
- By 2035, EVs in the US are expected to represent 9000 GWh of energy storage and 25,000 GW of inverter capacity, significantly more than the amount of grid storage or new generation that is currently planned. Further, ~1000 GW of behind-the-meter (BTM) distributed generation are also likely to be put in, which also potentially provides a "free" grid resource. These grid-edge resources can augment grid resiliency and reliability at low cost but require that behind-the-meter resources be integrated with grid operations, and further requires changes in current regulations and policies.
- For services requiring even higher levels of reliability than can be provided by a renewables-dominated grid, green hydrogen can provide the new generation resource needed. This can be done using existing gas generators that are modified to burn hydrogen, or using fuel cells that have higher efficiency.
- Finally, while zero carbon emissions by 2050 represents a great vision and target, it is very likely that there will be large sectors that may not be able to achieve the zero-carbon goals completely. This can include specific industry sectors such as general aviation. It can also occur because existing gas generation had to be ramped up because the sun did not shine, and the wind did not blow for an extended period of time. For such instances, a viable strategy could be to use technologies such as Direct Air Capture (DAC) to contemporaneously absorb the CO_2 that is generated (not in the distant future through an uncontrolled process such as planting trees). In such cases, the cost of fuel can include the cost of absorbing and permanently sequestering the equivalent amount of CO_2 that was emitted, where the extraction and sequestration of the CO_2 is measured and validated. Such a strategy allows flexibility and a path to continue to operate without getting bogged down in low-probability corner cases. It also provides a real business case for DAC that is not linked to a broad carbon tax.
- To soften the economic impact of such a transition to carbon-free operation (for applications as seen above), policies could be put in place to require that, say 10% of the CO_2 emissions must be absorbed starting in 2025, increasing to 100% by 2050—including some forward-leaning incentives in the early years. This will set a trajectory toward zero emissions that incentivizes both the aviation/energy industries and the DAC industry to improve on existing solutions, and which will trigger a strong competitive response.
- LCOE may not be the right cost metric for an energy ecosystem dominated by "free" but variable renewable energy resources. A better pricing mechanism is

needed that allows for recovery of capital investments, and which generates real-time pricing signals that allow dynamic and autonomous balancing of generation and load.

- The most basic human needs should be served at the highest level of resiliency and reliability (possibly as a human right). However, achieving higher levels of reliability and service level for non-critical loads (including commercial and industrial loads) can be more expensive. Such a need can be met either by the consumer through a behind-the-meter resource, or by the utility for a premium service charge.

While the above discussion focuses on the resources, this book has also discussed, often in nuanced detail, the challenges that the grid faces to integrate these new resources and operating requirements. Today, the grid is operated as a top-down, centralized system under the tight control of the grid operator, with rules that generators have to follow to ensure that the interconnected multi-generator system is stable. We have seen that this model can possibly continue for some time but is being increasingly challenged as the number of inverter-based resources connected to the grid increases.

The future grid needs to act like an ecosystem, an autonomous and secure marketplace that allows suppliers and consumers of electrical energy to securely trade energy and power on the grid platform, while also ensuring that the real-time-must-run (RTMR) physics-based requirements are addressed, and that the overall system remains stable. At the same time, every element connected to the grid ecosystem will need to act locally and in real-time to support and balance the grid, and to do it by following common rules (a model-based system that computationally determines fast-acting responses and relies on low-latency communications, tends to be fragile and unstable, and may be difficult to scale). The system needs to be built using flexible plug-and-play components that are modular, scalable, and can interoperate across technology generations and vendors. We have seen that this is possible today and can provide a viable building block for the future grid.

In this utopian 2040 future, materials will be processed using clean renewable energy. Buildings will become energy self-sufficient. Organic industrial feedstock materials will be made using carbon extracted from the air. Industrial processes, such as metal extraction, cement, fertilizers, desalination, and direct carbon capture will operate using clean energy, and will not generate carbon emissions. New appliances will be developed that are more energy efficient, including for cooking, heating, and cooling, allowing system optimization to manage the variability of the primary generation resource. The technologies that can enable such a future are already here, many in commercial use, some in the lab—ready to follow down a path of "innovation by design" to get to market along an accelerated timeline.

Research universities and national labs that receive significant funding from government agencies can play a significant role in how the right solutions are created and aligned for accelerated impact. Deep-domain disruptive technologies can be nurtured within the research centers and national labs where access to all tools and expertise is available, until the innovation risk is reduced, and it becomes clear whether the technology can be licensed to a major incumbent, is used as the basis of a startup, or is relegated to the shelf as interesting knowledge but a poor solution to solve real-life problems—all of which are valuable and acceptable outcomes. In the US, no other institution, certainly not industry, has the capability to address the deep-domain questions that often underlie major disruptions. But research labs must become proficient in, not only doing the science, but also in addressing the broad issues that can help accelerate time to impact.

Universities also play a critical role in training the scientists, engineers, practitioners, managers, financiers, and the entrepreneurs that will take these new disruptive technologies to market—but even there, the fast-moving technology cycles will necessitate a different model of lifelong learning.

Government also plays a critical role in guiding this transition forward. They can set the broad societal goals and provide the incentives and the guard rails that ascertain that all supported solutions follow the principles that will ensure that rapid scaling can occur—including interoperability, use of abundant materials, inclusion of life-cycle costs, and sustainability.

Given a clear shot at the end goal and the rules under which they can operate, entrepreneurs and investors are ready to take these new de-risked opportunities and turn them into reality. Once these de-risked technologies have been proven in terms of value delivered, there is no limit to how efficiently and quickly they can scale—if they are given the freedom along with the constraints under which they have to operate.

It is also critical to maintain a vigilant lookout for those who will abuse the public trust and pollute or damage the world again for short-term personal gain. Short-term incentives and subsidies can provide directionality toward actions, increasing the chance of achieving desired goals when the timing of the policy, state of technology readiness, and customer needs are aligned. Policies can deter those who will abuse the system or put burdens on society to clean up after them. They should be penalized and taxed, rather than be the beneficiaries of long-lasting politically driven subsidies. The first step is to identify and track key metrics and parameters with technologies such as IoT and satellite tracking of emissions—which is now feasible. The objective is not to infringe on a single person's rights, but to uphold the rights of millions, including the rights of future generations, ensuring that every generation restores the earth's ecosystem to how they received it, or pays the next generation to restore it!

The objective is not to infringe on a single person's rights, but to uphold the rights of millions, including the rights of future generations, ensuring that every generation restores the earth's ecosystem to how they received it, or pays the next generation to restore it!

We could soon have a world in peril, on the cusp of unpredictable and significantly damaging climate change caused by rapid and inefficient use of natural resource by humans. Without the exponential technologies that have been developed and are now beginning to be deployed at scale, we feel there would have been no chance for reaching a desirable end point on this journey that began 6000 years ago. We have seen that a strategy—solely based on penalties and taxes, that pits haves against have-nots, and that allows powerful incumbents to operate unfettered does not lead to desirable societal and environmental outcomes.

Today, we feel that we have one last chance to get things right. We may not cure everything, but we can put ourselves on the right path to reduce the level of climate adaptation and mitigation needed. These new exponential technologies give us the power to realize a world of sustainable, equitable, economic, reliable, resilient, and safe energy for all. This will help address the challenge of climate change, will create a more equitable and just distribution of energy and resources, will dramatically reduce anthropogenic carbon emissions, and will cut overall energy used by humankind by more than 50%. And that would be a good start!

References

1. Doerr, J. (2021). *Speed and scale: An action plan for solving our climate crisis now*. Portfolio/Penguin.
2. Divan, D. *AAAS annual meeting – 3rd plenary session*: https://www.youtube.com/watch?v=HW-KbSWQ5r4
3. https://www.bcg.com/publications/2021/deep-tech-innovation

Epilogue—Next Steps

The almost three years that it has taken us to write this book have also been accompanied by tumultuous and rapid global change. The world is moving back toward a new normal after the most debilitating days of the Covid-19 pandemic. Putin's invasion of Ukraine, along with the use of oil as a bargaining chip, has caused turmoil in the energy sector, particularly in Europe, and has pushed back many of the carbon goals that were being pursued. The Middle East is swirling with conflict, human anguish and forces that are fanning the winds of war. On the climate front, COP 27 (Cairo 2022) and COP 28 (Dubai 2023) have come and gone, with more handwringing as CO_2 levels continue to grow unabated, and governments continue to try to balance the objectives of economics and climate.

Yet, there are also many bright spots. EVs, PV solar, energy storage, and wind energy are surpassing previous records as adoption is accelerating in virtually every part of the world. The US has transitioned from the former Trump administration, which did not believe in climate change or renewable energy, to the current Biden administration, which has passed major legislation that significantly impacts the future of both—energy and climate. These new resources and legislation, including the $973B 2021 Bipartisan Infrastructure Law (BIL 2021), $1.1 T Inflation Reduction Act (IRA 2022), and the CHIPS and Science Act (2022), and recent cybersecurity measures, together commit over $555B for climate and the energy transition. Transportation, including automobiles, trucks, and aircraft, are looking at decarbonization on very aggressive timelines. US utilities are increasingly committing to decarbonization of their sector by 2050 and are gearing up to use the cost-share offered by new government initiatives to accelerate deployment of investments related to the energy transition. Venture capital and investors are funding unprecedented amounts to solve the really tough and complex problems in energy and climate.

So, have we reached a tipping point?

Is it now simply a matter of waiting for the outcomes to be reached and for our problems to be behind us?

We are concerned that may not be the case.

Given time, we are sure we would somehow muddle through, as we have done in the past, and get to a new transformed energy ecosystem. But our current crisis is

D. Divan, S. Sharma, *ENERGY 2040*,
https://doi.org/10.1007/978-3-031-49417-8

different—we simply do not have the time for a chaotic process of 50–100 years as politics, policy, technology, and economics collide to work out a new normal.

We are also concerned that there isn't a structured way, that is agreed upon by everyone, to deal with the rapidly evolving twenty-first-century exponential technologies, which will keep roiling our plans for well-designed processes and specific and predictable outcomes. And finally, we see new events, almost every month, that show the impact that climate change is already having, not only in terms of human suffering but also in terms of staggering economic impact.

As an invitee and an attendee at the White House Electrification Summit held on December 14, 2022, Deepak had a unique vantage point to see some of the fast-moving initiatives that will definitely shape how the next few years evolve in the world of energy and climate, especially in the US. It is clear that the Biden administration has a competent team leading the Department of Energy (DOE), Department of Transportation (DOT), and Office of Science and Technology Policy (OSTP), setting aggressive goals that involve decarbonization, electrification, economic development, self-reliance, equity, resilience, as well as a refocus on domestic manufacturing for the new technologies associated with the coming energy transition (particularly to reduce reliance on supply chains that can be influenced by autocratic regimes that do not always play by international rules). The focus was on immediate deployment, with many Funding Opportunity Announcements (FOAs) providing 50–80% cost share for targeted projects. The DOE Loan Program Office (LPO) could, in addition, provide loans of $100 M to $1B for deserving clean energy projects, helping advanced new technologies get to scale and commercial viability.

At the WH Summit, the case was made for 2000 GW (2× of current US generation capacity) of new clean energy by 2035 to support electrification and to ensure equity, and the need for new forms of innovation was also identified. The discussions focused on important concepts such as Democratization of Energy, a "collaborative" and transactive grid, resiliency, and equity for indigenous tribes and economically disadvantaged communities—themes we have touched on in this book also. The question of a new business model for electricity supply was raised, driven by the emergence of zero-marginal-cost non-dispatchable resources (e.g., PV solar) that would dominate the future grid. Cybersecurity, physical security of bulk-grid assets (such as substations and large power transformers), and the growing importance and possible proliferation of microgrids as a resiliency solution was extensively discussed. Many considered the availability of $5B to drive the installation of 500,000 EV chargers to be a key enabler for electric vehicles.

But our concern, as we have detailed in this book, remains about whether these grants and subsidies will move the right technologies and solutions forward, and do it at the scale needed over the next decade.

At the WH Summit, there seemed to be little discussion about the challenges posed by an emerging new paradigm, or the technologies with steep and sustained price declines that could shake the financial assumptions that were the foundation for the proposed projects. There was an assumption that the technologies needed to achieve the objectives they proposed were already available, and that the major vendors were ready (and happy) to deploy solutions they already had.

The focus for the major Investor Owned Utilities (IOUs) who were present was not on integration of these fast-moving technologies with their systems, but on technologies such as dynamic line rating, fault identification, and advanced metering—existing mature technologies with marginal impact that they have been implementing for the last 15 years.

The need for massive electrification was recognized, but the challenge of growing the infrastructure by 3× was not explicitly addressed. There seemed to be an emphasis on "digital" technology for data sharing, analytics, and cybersecurity—but again, only within the existing centralized grid paradigm, with little acknowledgment of technology challenges that massively distributed and decentralized real-time-must-run systems (with dynamic control, coordination, and real-time-must-run or RTMR requirements) need to overcome. There seemed to be little discussion about the need for new solutions, the disruptions, and the new paradigms that may be needed to achieve the very goals that were being laid out.

It is clear that the billions of dollars in subsidies and grants available will be spent, and will hopefully move the ball forward, but it is unlikely that they will be used to implement the more advanced solutions truly needed to shift our existing paradigm.

Looking at the continuing slew of announcements about major new initiatives on the energy transition, and the progress being made globally, we were also concerned about whether we ourselves were being a little paranoid. Maybe things were really under control and would settle down soon into a new "normal." For instance, FERC's recently issued Order 2023 addresses the need to shorten the interconnection queue for new renewable projects and directed utilities to prefer non-traditional solutions that were lower cost and faster to deploy. Similarly, new initiatives to build transmission, and required storage on the grid, are being announced and are also to be welcomed.

On the other hand, many conversations over the last few months with senior executives in government agencies, industry organizations, grid operators, utilities and manufacturers who supply to the electricity and EV infrastructure have reconfirmed that, while they are excited about the changes underway, they also have elevated concerns that many key problems related to technology, business models, and policy that would be manifested as the system scales had not been addressed yet. Concerns about a severe shortage of a workforce trained in the new technologies, and the issue of retraining of the existing workforce were also top of mind.

A recent trilateral workshop in Germany that was organized by the National Academy of Sciences in the USA, Germany, and Israel reaffirmed the urgent need for action and showcased exciting new technologies. Yet, everyone seemed to think of the grid as a "resource," available to accept and to deliver energy wherever and whenever someone needed it. The questions of distributed and decentralized control were not part of the conversation, and questions of fast physics coupled systems were not even recognized as being an essential part of the discussion. For the German contingent, with strong industry representation, there was a recognition that much needed to be done, yet their collaboration processes were slow and ponderous, raising the question of achieving the desired scale in the time available.

There was much discussion in the US and Israeli contingents on the role that entrepreneurs played, and the need to accelerate science to impact—topics we have discussed ad nauseum in this book.

Again, we want to reemphasize that the question is not about whether the energy transition will occur—it will, because it is driven ultimately by economics. The challenge, as we have discussed in this book, is whether we can anticipate major issues, manage their impact, and accelerate the transition—because that would be the right economic decision, and could also mitigate some of the impact of ongoing climate change. Can this new alignment, between what have traditionally been two divergent groups, be the start of a conversation that brings together the best minds globally to find a path forward?

As we mull next steps, one question haunts us. Is this the best (the collective) WE could have done? Have WE sufficiently derisked the possibility of a backlash against the energy transition, that can slow us down, increase inequity and increase cost? Have WE missed a golden opportunity to meet the timeline for decarbonization, while forging a path that is also economically viable and leads to global energy sufficiency? Do the new technologies and insight provide us the tools to create a roadmap for a new utopian future? While we are optimistic and remain hopeful, time alone will show what kind of world we will leave behind for our children and grandchildren, and whether we will have acted to not only make our own lives better, but to also preserve the rights of future generations.

Bibliography

1. Atkinson, R. C., & Blanpied, W. A. (2008). Research universities: Core of the US science and technology system. *Technology in Society, 30*, 30–48.
2. Bakke, G. (2016). *The Grid – The fraying wires between Americans and our energy future*. Bloomsbury Publishing Plc.
3. Bakke, G. (2017). *The grid, 'The fraying wires, between Americans and our energy future'*. Bloomsbury.
4. Brown, M. A., & Sovacool, B. K. (2011). *Climate change and global energy security: Technology and policy options*. MIT Press.
5. Brown William, H., Malveau Raphael, C., McCormick, I. I. I., Hays, W., & Mowbray, T. J. (1998). *"Anti patterns" – Refactoring projects in crisis*. Wiley.
6. Bush, V. (1946). *Science – The endless frontier: A report to the president by Vannevar Bush, Director of the Office of Scientific Research and Development, July 1945*. United States Government Printing Office.
7. Carney, M. (2015, September 29). *Breaking the tragedy of the horizon – Climate change and financial stability*. Lloyd's of London.
8. Chesbrough, H. W. (2006). *Open innovation*. HBS Press.
9. Christensen, C. (2012). *The Innovator's dilemma*. Harvard Business Review Press.
10. Clark, I. I., Woodrow, W., & Cooke, G. (2016). *Smart green cities, towards a carbon neutral world, A Grover Book*. Rutledge Group.
11. CLIMATE – Lapham's Quarterly, Volume XII, Number 4, Fall 2019.
12. Cohn, J. A. (2017). *The grid: Biography of an American technology*. MIT Press.
13. Costello, K. (2016, May). *A primer on R&D in the energy utility sector* (Report no. 16-05). National Regulatory Research Institute.
14. Divan, D. AAAS annual meeting – 3rd plenary session: https://www.youtube.com/watch?v=HW-KbSWQ5r4.
15. Doerr, J. (2021). *Speed and scale: An action plan for solving our climate crisis now*. Portfolio/Penguin.
16. Drucker, P. F. (1993). *Innovation and entrepreneurship*. Harper Collins Edition. First published in 1956.
17. Drucker, P. F. (2001). *"Innovation and entrepreneurship" first published in 1956, and later gain in 1992 by Collins. Also, "The essential Drucker"*. Harper Business.
18. Drucker, P. F. "The discipline of innovation" on innovation, HBR's 10 must reads.
19. Dyer, J., Gregersen, H., & Christensen, C. M. (2011). *THE Innovator's DNA – Mastering the five skills of disruptive innovators*. Harvard Business Review Press.
20. Enkhardt, S. (2022, August 1). *Germany deployed 3.8 GW of PV in first half of 2022*. PV Magazine.
21. Fertina, N. (2022, August 26). *3D-printed solar cells are cheaper, easier to produce, and deployable at speed*.

D. Divan, S. Sharma, *ENERGY 2040*,
https://doi.org/10.1007/978-3-031-49417-8

22. Gryta, T., & Mann, T. (2020). *Lights out: Pride, delusion, and the fall of general electric.* Houghton Mifflin Harcourt.
23. Gryta, T., & Mann, T. (2020). *Lights out; 'Pride, delusion and the fall of general electric'.* Mariner Books.
24. Haggerty, J. (2020). *Sunny places could see average solar prices $0.01 or $0.02 per kilowatt-hour within 15 years.* PV Magazine.
25. Hamilton, J. D. (2010, December 22). Historical oil shocks. In R. E. Parker & R. Whaples (Eds.), *Routledge handbook of major events in economic history* (pp. 239–264). Routledge.
26. https://www.bcg.com/publications/2021/deep-tech-innovation
27. Interesting Engineering. https://interestingengineering.com/innovation/3d-printed-solar-cells-are-cheaper-easier-to-produce-and-deployable-at-speed
28. International Energy Agency. (2004, October). *World energy outlook 2004* (Flagship report). IEA.
29. Jafee, A. M. (2021). *Energy's digital future: Harnessing innovation for American resilience and National Security.* Columbia Press.
30. Jones, J. (2006). *Empires of light, 'Edison, Tesla, Westinghouse, and the race to electrify the world'.* Random House.
31. Malik, N. (2022, August 30). *Negative power prices? Blame the US grid for stranding renewable energy.* Bloomberg.
32. National Academy of Engineering. (2015). *"Frontiers of engineering" – Reports of leading-edge technologies impacting upcoming decades*
33. Nussey, B. (2021). *Freeing energy: How innovators are using local-scale solar and batteries to disrupt the global energy industry from the outside in.* Mountain Ambler Publishing.
34. Olah, G. A., Goeppett, A., & Surya Prakash, G. K. (2005). *Beyond oil & gas: The methanol economy.* Wiley – VCH.
35. Panicker, N. R. (2021). *An entrepreneur's journey – The joy of dreaming.*
36. Pelton, J. N., & Singh, I. B. (2015). *Digital defense.* Springer.
37. PitchBook. (2022, September 8). *Carbon & emissions tech launch report* (Report). PitchBook.
38. Shahan, Z. (2017). *IEA gets hilariously slammed for obsessively inaccurate renewables energy forecasts.* Clean Technica.
39. Sharma, S. (2014). *Energy: India, China, America and Rest of the World* (Chapter 8).
40. Sharma, S. K. (2014). *"Energy: India, China, America and rest of the world", The 3rd American dream.* Creative Publications, Amazon.
41. Sharma, S. K., & Meyer, K. E. (2019, April). *Industrializing innovation – The next frontier.* Springer Nature.
42. Sharma, S., & Meyer, K. (2019). *Industrializing innovation – The next revolution.* Springer Nature.
43. Shrier, D., & Pentland, A. (2016). *Frontiers of financial technology.* Visionary Future.
44. Siota, J. (2018). *"Commercializing discoveries at research centers" – Linked innovation.* Palgrave Pivot MacMillan.
45. Sivaram, V. (2018). *Taming the Sun: Innovations to harness solar energy and power the planet.* MIT Press.
46. Smil, V. (2017). *Energy and civilization: A history.* MIT Press.
47. STEM Connector, & James, M. (2015). *"Advancing a jobs-driven economy" – Higher education and business partnerships lead the way.* The Entrepreneurial Publisher.
48. Stevens, P. (2022, September 8). *Solar installations will triple in 2027 thanks to climate bill, report predicts.* CNBC.
49. Thomas, H., Lorange, P., & Sheth, J. (2013). *The business school in the 21st century.* Cambridge University Press.
50. Weise, E. (2022, June 24). *A 'Wow' moment: US renewable energy hit record 28% in April. What's driving the change?* USA Today.
51. Willuhn, M. (2022, July 7). *Germany raises feed-in tariffs for solar up to 750kw.* PV Magazine.
52. Yergin, D. (2011). *The quest: Energy, security, and the remaking of the modern world.* Penguin.

Index

D. Divan, S. Sharma, *ENERGY 2040*,
https://doi.org/10.1007/978-3-031-49417-8